# THE DEEP

# About the Author

**Alex Rogers** is Science Director of the Norwegian foundation REV Ocean, a Visiting Professor at the Department of Zoology, University of Oxford and a Senior Research Fellow of Somerville College, University of Oxford. As one of the leading marine biologists in the world, Alex advises the UN, Greenpeace, the WWF, and the G8 countries on ocean ecology, and recently served as a scientific consultant on the BBC's *Blue Planet II* series.

# THE DEEP

## THE HIDDEN WONDERS
## OF OUR OCEANS AND HOW
## WE CAN PROTECT THEM

ALEX ROGERS

WILDFIRE

First published in 2019
by WILDFIRE
an imprint of HEADLINE PUBLISHING GROUP

1

Cataloguing in Publication Data is available from the British Library

Hardback ISBN 978 1 4722 5395 8
Trade paperback ISBN 978 1 4722 5394 1

Typeset in Adobe Garamond by CC Book Production

Printed and bound in Great Britain by
Clays Ltd, Elcograf S.p.A.

*'In the darkening twilight I saw a lone star hover,*
*gem-like above the bay.'*

These were the last words written by Ernest Shackleton in his diary on 4 January, 1922, penned onboard the expedition ship *Quest* in Grytviken harbour, South Georgia. Hours later he died of a heart attack. His resting place above the bay is a shrine to explorers the world over.

*To Candida, Freya and Zoe,*
*my love for you is as endless as the ocean.*

# Layers of the Ocean

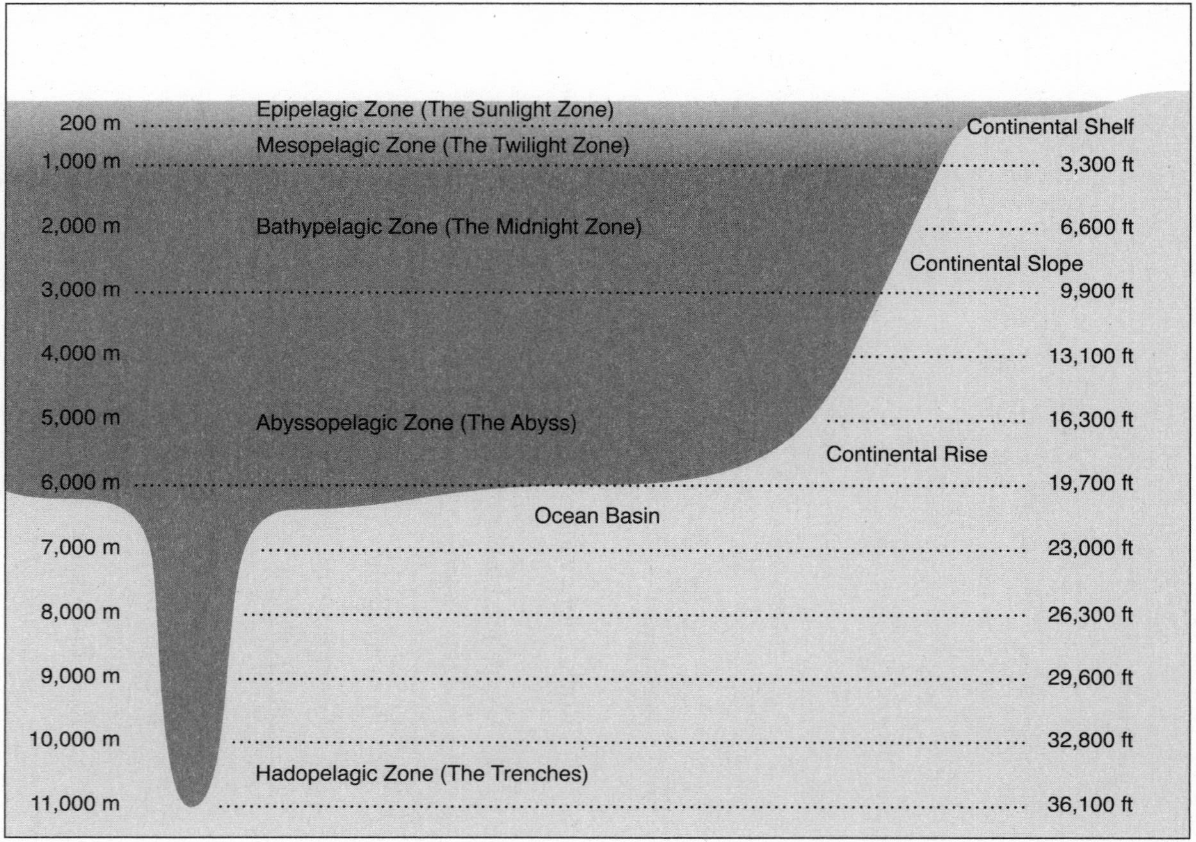

Epipelagic Zone (The Sunlight Zone)

Mesopelagic Zone (The Twilight Zone)

Bathypelagic Zone (The Midnight Zone)

Abyssopelagic Zone (The Abyss)

Hadopelagic Zone (The Trenches)

Continental Shelf

Continental Slope

Continental Rise

Ocean Basin

- 200 m
- 1,000 m — 3,300 ft
- 2,000 m — 6,600 ft
- 3,000 m — 9,900 ft
- 4,000 m — 13,100 ft
- 5,000 m — 16,300 ft
- 6,000 m — 19,700 ft
- 7,000 m — 23,000 ft
- 8,000 m — 26,300 ft
- 9,000 m — 29,600 ft
- 10,000 m — 32,800 ft
- 11,000 m — 36,100 ft

# Contents

# Prologue

Fewer people have been to the deepest part of the ocean than have been to the moon. Only three people have visited the Challenger Deep, which forms part of the Marianas Trench in the western Pacific Ocean. Twelve astronauts have visited the moon. Why this should be so, is complicated. Space exploration has always been associated with wonder and awe. There is something heroic about it that captures the imagination – a dangerous endeavour. It is humankind pushing on the final frontier, trying to understand our place in the universe. Then there is the exciting technology and unique imagery: the futuristic white spacesuits with their mirror-finish visors, enormous rockets and now, even space planes. It was the Apollo 17 mission in 1972, the last one to place men on the moon, that took the iconic 'blue marble' image of the Earth from space, graphically demonstrating how the surface of our planet is mostly covered by ocean.

In the time of the naval explorers Ferdinand Magellan,

# The Deep

Christopher Columbus and, later, James Cook, I would imagine that exploring the ocean was equally as exciting, and certainly as challenging, as space travel is now. However, we have somehow lost sight of what lies beneath the ocean surface and why it matters. Even now, the vast majority of this wilderness – which forms the largest ecosystem on Earth – has never been seen by human eyes, let alone explored or investigated by scientists. When most people look out onto the ocean, perhaps from one of our coastal cities or on a cruise ship, all they will see is the ocean surface. The more adventurous might go for a swim or snorkel and see the pretty fish that live in a few feet of water. Some even scuba dive into the surface layer, the skin of the ocean. It is easy to forget the ocean is nearly 11,000 metres deep at the Marianas Trench and has an average depth of 4,200 metres. Mount Everest could be submerged comfortably in the Challenger Deep. So the ocean is remote, and not only does it slip past our perception, but it has even become associated with bad and frightening news, whether it be overfishing, pollution or the insidious effects of climate change. Such is the enormity of many of these problems that some of us simply shut it out. We do not want to look down into the depths at the disaster that is unfolding at our own hands, but rather up into space and its endless wonders. Perhaps, as some claim, space represents the escape route from our ruined planet. They think that there is another chance to set up home elsewhere. But I will always believe that we need to be looking at our inner space instead.

I have spent the last 30 years trying to understand how life is distributed in the ocean. In 2016 I found myself diving with the Nekton Foundation on Plantagenet Bank, a seamount – undersea mountain – lying just over 40 kilometres to the south of Bermuda. This dormant volcano rose from a depth of more than 2,000 metres at its base, to a summit just over 40 metres below the surface. In its time the seamount had played host to a Cold War submarine listening station. The remains of a steel tower built atop the summit lie underwater, still linked to an array of cables, originally connected to sound detectors, fanning like a web down the steep sides of the volcano. The tower was called Argus, and thus the seamount acquired the new name of Argus Bank. Nowadays it is a destination for sport-fishing boats operating offshore from Bermuda hunting for marlin, tuna and other ocean game.

That day, I must admit, I climbed into the submersible with a few butterflies in my stomach. *Nemo* was a Triton 1000/2 and was unusual in that the entire pressure hull was made from acrylic. Imagine diving in a giant goldfish bowl capable of taking you to 300 metres depth and able to move in all three dimensions completely untethered from the ship. The visibility for the bathynauts (submariners that dive below 200 metres) from this new design of submersible was spectacular, nearly 360 degrees as well as above your head and down between your feet. The disconcerting thing was that, of course, the fish were looking in, not out, and sometimes passed extremely close, appearing suddenly over the shoulder or from overhead. Earlier, using a fast launch with a sonar, we had discovered that the nautical charts for Plantagenet Bank were wrong. The trace on the echo sounder screen also indicated that from a depth of 70 metres the seamount flanks plunged near-vertically for hundreds of metres. It had taken us some time to find a place where there was a narrow ledge at 200 metres depth. A submersible should always have seabed below it that is at or shallower than its diving limit, in case there is a power failure and the vessel sinks. Exceeding the pressure of the sphere could lead to it imploding and pulverizing those inside. There was a large safety margin in terms of the maximum depth of operation, but even so, I tried not to think too deeply about the possibilities of an accident.

Robert, our pilot, was clearly excited at diving Argus for the first time. The hatch of the sphere was closed and locked from inside, and then the pre-dive checks run through. As the sun was up, the temperature inside the sphere rapidly began to climb, so as soon as the team outside had finished sealing in the cameras for survey work we were lifted off the deck. The submersible was hoisted over the end of the ship underneath a large inverted U-shaped frame (called an A-frame), and as we splashed into the water we were met by an inflatable with a 'swimmer', the person who unhooks the submersible from the lifting straps. Finally, we were ready to dive and Shane, the Dive Operations Officer, called over the radio in his southern American accent, 'Ready to dive . . . Dive, dive, dive.' The submersible lurched forward and down. Blue water rose over the front face of the acrylic sphere giving a real sensation of falling into the depths. My stomach heaved, both as a result of the sphere diving head first and with nerves. Not quite everything was strapped down and a pencil and water bottle fell around my feet. Robert evened *Nemo* out just under the surface which, looking up, resembled turbulent mercury. We vented air with a sound like a bath draining and down we went.

# Prologue

The first thing that strikes you is the sudden silence. Inside, there is the quiet hiss from fans and the odd whirring from the thrusters, but on descent it is mostly peaceful. We looked down between our feet as the submersible sank under its own weight. The seamount summit materialized at a depth of 60–70 metres, and, to my astonishment, I saw that it was carpeted in green weed resembling torn rags. The seaweed, a pale-green species called *Sporochnus*, was lying flat and rippling as strong currents poured over the seamount. Underwater mountains act as huge obstructions to passing water, so it pushes around, up and over them, accelerating as it goes. This is one of the reasons seamounts host such spectacular growths of life, as the strong currents bring nutrients for algae and particles of food for animals, including plankton, the tiny organisms living near the ocean surface. On the seamount summit it was like a busy city. Fish were balancing in the current by rowing with their pectoral fins and the odd flicking of their tails, diving to snap up a morsel of food before submerging below the surface of seaweed. There were coneys, small groupers with a white belly and crimson back, and tiny *Chromis*, a bright-blue fish with a yellow tail, careering in clouds. A red hind, a larger grouper with big lips and a pale-grey body speckled with red spots, peered out from the seaweed. Pale-yellow, disc-shaped butterfly fish with black bars chased each other, defending territories or nesting sites. An eerie short bigeye hovered, a narrow but broad-bodied silver fish with large jaws slung under huge eyes, very obviously adapted to hunt in the darkness. Particles of what looked like snow zipped past the submersible, along with clumps of seaweed. Robert battled to keep *Nemo* stationary while we waited for the second submersible, *Nomad*, but the current buffeted us left and right like a kite in a gale. We therefore settled onto the summit and had a bite to eat – peanuts, in my case. Not many people can claim to have had lunch sitting on a seamount.

After another 10 minutes we decided we were wasting time and burning batteries, so we called *Nomad* over the radio and told them we were heading for the edge of the summit; the drop-off. An enormous barracuda suddenly materialized beside us, body like a medieval broadsword with dark-blue bands. It eyed us, large pointed teeth jutting at different angles in a mouth meant for murder. Sleek and deadly, it was an amazing sight as we slipped over the summit edge, which felt like stepping off the ends of the Earth into an infinity of blue. The drop, down a near-vertical limestone cliff, was vertiginous, but as the echo sounder on the survey boat had indicated, the cliff turned into a steep slope at around 100 metres depth. We dropped down to 200 metres depth and settled the submersible to begin surveying the animal life on the flanks of the seamount.

Although I was concentrating on maintaining a steady survey line at a constant distance from the seabed it was hard not to be distracted. The scene before us was beautiful. The steep slope was formed by pale-green limestone bedrock. Sporadically we encountered limestone ridges and strangely eroded features including a sculptured, twisted rock arch. All was draped in a scattering of white sediment reminiscent of snow. In places twisted wire coral, two metres high, carpeted the rock, especially where it formed lips and crests. Pink and orange brittle stars could be seen clinging to the corals, which were a feathery white. Now and again we would spot a sea urchin with fine, long, curved needle-like spines. Their shells, or tests, had a stellate black-and-white colour pattern, and when approached by the submersible they moved rapidly across the rock frantically waving their spines. Where the rocky terrain was interrupted by holes, pinnacles or other irregular features, shoals of small pink fish, up to 7–8 centimetres long with yellow backs and tails, moved away from the submersible in short bursts. These were sometimes chased by small saddle bass, immediately recognizable by dark vertical bars and a fluorescent pale-gold vertical stripe on their bellies. Fish usually hidden in the gloom were revealed by *Nemo*'s lights and attracted these mini predators, which were always in the vicinity of the pinks (we were uncertain of the identity of these fish). It was quite a surprise to come across a large yellow moray eel, mottled with white, and perhaps a metre and half long, probing the nooks and crannies of the limestone and searching among scattered cannonball-like rocks called rhodoliths. These are limestone concretions formed by red algae in shallower water which tumbled down the slopes of the seamount along erosional gullies. We watched the eel for several minutes. It did not seem concerned about the presence of a brightly lit submersible. Later I found out that the species had not been recorded in Bermuda before, but there were sporadic records of it around islands in the North Atlantic. It had originally been described off Madeira.

But there was a point during the dive that I will never forget, something that conveyed to me the enormity of the task ahead for humankind. After doing the next survey we turned, and I spotted the other submersible, *Nomad*, in the distance, the size of a toy, lighting up a vast underwater cliff in pale blue surrounded by deep, dark blue-black. In this moment, reality blurred. I suddenly had a vision of being in a small spacecraft, watching another ship manoeuvring around the steep face of a giant asteroid or some alien planet. My breath caught. What a sight that was, floating in the deep blue; a bubble of excitement rose in my chest. I had seen many extraordinary things in my career, but that is a picture that is burnt on my memory forever. That moment, for me, captured the vastness of the ocean, our inner space. We were exploring an environment that would be instantly fatal to human life were it not for the technology of the plastic bubble and gadgetry we were surrounded by. We were, indeed, engaged in a heroic endeavour, to explore the ocean, to boldly go where no human had been before. Only by doing this could we reveal the beauty and importance of the deep to our fellow humans and show them graphically the damage being done to the planet.

My love affair with the ocean began many years ago as a young boy spending the summers fishing with my grandfather or scouring the rock pools of a quiet corner of the coast of County Sligo in Ireland for marine creatures. That thrill of hunting in the sea has never left me, though now it is the hunt for new habitats or species rather than the hunt for dinner. My job has taken me from the stormy waters of the North Atlantic to the remote southern Indian Ocean and Antarctica. It is exciting work, full of mystery, surprises and even, should you look for it, danger. The deep sea is not only our largest ecosystem, but is also the least explored. It has been estimated that only 0.0001% of the deep-sea floor has been sampled by scientists. Thus, the largest ecosystem on Earth, containing 1.3 billion cubic kilometres of water and covering an area of over 360 million square kilometres, is the least known. It is remarkable that we have better maps of the moon and of Mars than the bottom of the ocean. But this is for the simple reason that these bodies are visible to orbiting satellites, whereas the bottom of the ocean is covered in water and is largely invisible to satellite sensors below a few tens of metres, even in the clearest seas. While most of this vast wilderness remains unmapped, it contains perhaps 90% of all life, and the physical and biological processes within it are critical to supporting life on Earth.

In this book, I hope to show you how discoveries in the deep sea have changed our perceptions of how life may have originated and the extreme conditions in which it can thrive, and, ultimately, of our place in the universe. I will also show you how, despite the enormity of the ocean, we are damaging it catastrophically through over-exploitation of its resources, pollution and the insidious and global effects of climate change. It is a topic that is ever-present in my mind, like something permanently in my peripheral vision, and when I concentrate on it I am afraid – particularly for my children. Sometimes this damage is caused by our own carelessness, but often it is deliberate, motivated by profit and an economic system that places no value on the very foundation of our existence – nature. There is no second planet Earth for the foreseeable future so changing the ocean's current trajectory from one of degradation to recovery is critical for the continued existence of life on Earth as we know it.

# The Deep

There have been many doom and gloom stories about the ocean. While this book paints an honest picture of the dangers faced by marine animals and ecosystems, I believe that we still have everything to play for. By taking you, the reader, into the deep sea, I hope to show you some of the wonders of our oceans: underwater mountains and hot springs, sea creatures of all kinds and dazzling coral gardens. I hope that by taking this journey, you will come to know a new world that is novel, strange and entirely unique – so much so that you will want to treasure it and participate in its care for future generations who, like us, will depend upon it.

# 1

# Rock Pools:

## *How I Became Obsessed with the Ocean*

My journey in to the deep sea began with my experiences as a child. I cannot overstate the importance of exposure to the natural world, whether this be a local meadow, the edge of the sea or even through media, such as books and television. As a child I was surrounded by people who were willing to listen, and to share my enthusiasm for fish, sea slugs, ants, butterflies or dinosaurs. My natural curiosity was ignited. This was further stimulated by walks in the local woods, trips to the zoo and, later, abroad. But it was also as simple as sitting around the television sharing a magical piece of film about sharks, coral reefs or a strange creature in a tropical rainforest. I began to appreciate the natural world as something to be valued and a topic on which I could communicate on equal terms with adults and share with them my fascination. My journey towards a career as a marine biologist, however, began during summer holidays in Ireland in the 1970s, more than 40 years before the submersible dives in the Sargasso Sea off Bermuda.

My grandfather had been a fisherman all his life and his boat was an open wooden vessel, coated in flaking white paint and about 12 feet long. It was tied up at the dock at Cloonagh, just along the southern shore from my grandparents' cottage on the coast of County Sligo in Ireland. Dock was a bit of misnomer for what was a very crude stone pier welded against the elements over the years with additions of concrete applied after the winter storms had taken their toll. It was four to five feet high and extended just a little way out to sea, meaning that the boat was often stranded by the receding tide on the surrounding limestone flags. I remember a particular day very clearly. It began, as with all these trips, as I watched the men – my grandfather, father and Uncle John – haul the heavy boat in fits and starts down the rocks and into the water. It was a fine sunny day in the bay and the ocean was a deep royal blue with gentle swells sometimes capped by lazy low breakers. I was always excited by the prospects of our voyage, as we pushed away from the land. What animals would I see? Would we get the chance to do some fishing?

Looking out from my seat near the bows of the boat the ocean stretched to the horizon. To the north-west lay Inishmurray Island, a low-lying limestone outcrop, framed by the distant cliffs of Donegal. We had visited the island previously as it had once been home to a small community, including some of my ancestors. It also boasted a well-preserved ancient drystone walled monastery dating back to the sixth century and the early days of Christianity in Ireland. There was a light breeze jostling my hair as the boat growled through the sea keeping away the stink of the bucket of bait, oil and the filthy brine below the boards that made up the deck. I have never been seasick, so the smells and the movement weren't an issue for me, even as a young boy. Fulmars, white and grey seabirds with short beaks, swept past, nearly dipping the tips of their long, narrow wings into the water. Their aerobatics, wheeling constantly over the waves, were mesmerizing to me as a young boy, and even to this day I count them among my favourite seabirds.

We passed along a coast formed of limestone strata tilted at varying angles into the sea, waving to the odd man fishing from the rocks, my grandfather or uncle invariably commenting on who he was, where he lived and what he might be fishing for. There was also the occasional other fishing boat, similar to my grandfather's, but some bigger, with a cab for the captain. A wave and occasional greetings were exchanged, but there was also an undercurrent of guarded suspicion and competition between all those earning their wage from the ocean. Eventually we reached the first of the tows of lobster pots marked by a soft plastic buoy bobbing at the surface. The lobster pots, or creels, had a wooden base weighted by concrete, and hoops of wood or plastic over which was stretched coarse blue or orange netting, with an entry tunnel for lobsters or crabs. A hole in the top of the creel was tied with string so bait could be placed inside between two cords and held with a slip knot. The bait was usually a partially decomposed piece of salted fish; mackerel, gurnard or sometimes an unfortunate crab, smashed on the gunwales of the boat and suspended in the pot still feebly curling its legs. The creels were tied in lines of a dozen or so to form a tow, strung out along the seabed.

The buoy was retrieved with a boat hook, and my father threw it onto the deck of the boat along with a bundle of rope covered in slimy brown algae, splashing water everywhere. I tried to avoid getting wet as the water was cold, even in the middle of summer. Then the hard work began as my father pulled the rope over the gunwale, hand over hand, with the weight of 12 pots and the rope dragging along the seabed. For me the anticipation of what might come up was almost unbearable. I leaned over the side of the boat to see the first pot materialize from the inky blue-black depths and whether it contained lobsters or any other creatures. The first pot was empty, as was the second and third, causing some frowns and growl-like mutterings. My uncle untied each one, threw any remaining rotten bait over the side and slipped in a fresh filleted gurnard before tying the creel up expertly and then stacking it in the middle of the boat. As the next pot materialized and was drawn to the surface I could see something in it. Buzzing with excitement, I started shouting: 'There's something there, I can see it moving!' The pot was heaved on deck, and to my delight it contained two snapping lobsters, tails flapping, spraying water and jerking around the bottom of the creel.

The lobsters were the European variety, magnificent crustaceans heavily armoured with a royal-blue shell paling to yellow with whitish spots on the underside. As soon as the water had drained from the creel, the lobsters backed into opposite corners wedging themselves between the mesh work. Their front end was protected by an armoured shield covering the head. They had a pair of blue-black eyes on stalks and long red antennae, which curved out from the head. Under this were long jointed legs ending in small pincers covered in short orange hairs. The back end was formed of a long segmented tail with a fan of plates at the end. They were armed with one large crushing claw with lumps and nodules on it and a lighter scissor claw with finely serrated edges. They held these up to catch an unwary finger as my uncle retrieved them from the pot. He lifted them from just behind the head where they could not reach to deliver a nip. On a previous trip I had seen my father do this and be caught by one of the claws causing him to fling his arm back, not realizing the lobster was still attached, launching the creature spinning through the air and into the sea with a splash to its freedom. Luckily, there had already been a debate as to whether *that* particular lobster was too small to keep.

My uncle held each lobster in turn on his lap between the legs of his yellow oilskins while he bound their claws with rubber bands to prevent them from fighting and damaging each other. One of the problems with culturing lobsters or reseeding lobster habitat with young animals is that they are terrifically aggressive and highly territorial and will kill and eat each other to occupy the best holes and crevices on the seabed. The lobsters were then thrown into a red plastic fish box lined with old sacking soaked in seawater to keep them damp until they could be stored in a tethered slimy old wooden box which was floating just off the shore from the pier. My uncle had shown me how to hold the animal and I very carefully picked one of them up to examine it while the next creel was being hauled.

As a child, staring into the face of a lobster, I was always struck by just how alien these creatures are. Lobsters cannot form expressions so it is somewhat like looking into the face of a fully armoured medieval knight with no clue as to what is underneath. The eyes are compound, like those of an insect, but the facets formed by the individual light-guiding crystalline structures which make up these eyes are not so obvious. As I was to learn many years later the lobster has a superposition compound eye, superbly adapted to perceive movement in dim light but unable to form sharp images. Instead, lobsters rely mostly on their senses of touch and smell to locate prey, their enemies and other lobsters. Projecting from below a thick barbed spine that juts from between the eyes was the pair of long antennae, deep red in colour and articulated at the base, used by the lobster to feel around its environment. Smaller pairs of antennae tucked between the larger are used to 'smell' chemicals in seawater. Additionally, the lobster is covered in tiny hairs, particularly located on the front legs and their terminal pincers, also used to taste the water. They can literally smell with their feet. Unsurprisingly, the animals are nocturnal, remaining in rocky or sandy holes during the day and leaving their shelters at night to feed on molluscs (snails, clams, mussels, etc.), crustaceans (crabs, other lobsters!), urchins, starfish and even grazing on seaweed or scavenging on dead animals. It was for this reason that the creels had to be left to 'soak' overnight, the juices leaking from the bait attracting the lobsters while they were on their nightly wanderings. Like we might stroll down the street at night and smell fish and chips before homing in on the chip shop, so too do the lobsters find themselves drawn to appetizing wafts from the reeking bait.

Still mesmerized by the creature I was seeing up close, I saw that the mouthparts of my specimen were knitting a mass of mucous bubbles. They were formed by several pairs of appendages, some flattened and some longer. These taste, manipulate and move food to a pair of vertically orientated jaws, which cut or tear pieces of food from their prey. The claws are too large and positioned too far forward on the body to pass food to the mouth and are instead used to kill or injure prey and to crush it. Great care had to be taken to turn the lobster over as the apparently floppy claws could suddenly grasp any loose clothing or exposed skin. Beneath, the articulated legs were fixed to the front half of the body through flexible membranous joints. The long segmented tail was also covered in soft membrane and held pairs of small paddle-like swimmerets.

We sometimes caught female lobsters with tens of thousands of small, black, grape-like eggs held in a massive cluster under the abdomen, and these would be returned to the sea to help conserve the stock. The eggs hatch as tiny larvae, almost invisible to the naked eye, which live in the plankton for 5–10 weeks before settling on the seabed. Many of the larvae are eaten by fish and other animals, and few make it to the juvenile stage living on the seabed. Little is known about the time between planktonic larvae and adult, but it is assumed the lobsters live in small burrows preying on smaller marine animals, such as worms. That this is still a mystery when we have been fishing lobsters probably for thousands of years, and they have been subject to intense scientific scrutiny, is a demonstration of how elusive knowledge of marine life can be. It is a theme I will return to time and again in this book. I like to imagine the young lobsters are on a self-taught apprenticeship, figuring life out as they grow during this hazardous time in their lives when a mistake can mean ending up as a meal for someone else, even a brother or a sister. There's something admirable about their independent and pugnacious nature, and this is a part of the reason lobsters are still fascinating to me today.

The spell having been broken and the first lobsters landed, the atmosphere changed and my uncle, my mother's brother, started smiling and joking with my father. They had met as young men because my mother's entire family migrated annually to Walthamstow in London to work during the winter in the building trade and this was where my father was living at the time. During this season my grandmother cooked for a local company in London. I have a distinct memory of a hall scattered with melamine tables with her dressed in a blue overcoat and defiantly in charge of the enormous steaming pots boiling away on stoves. She had a shock of red hair, a ruddy complexion and was the archetypal Irish matriarch: full of pride, opinionated and watchful over everybody's business.

After the first tow of lobster pots was hauled, my father and uncle lit up cigars producing clouds of woody smoke whipped away by the breeze as my grandfather sat by the engine looking ahead with intense concentration as he repositioned the boat. He had fished these waters since childhood, rowing out with my great grandfather, who by all accounts had been a very intimidating character, and a man not to be messed with. On my grandfather's word, the buoy, rope and first pot were thrown overboard, and as we motored along, the next pot would be hefted onto the gunwale to be released as the rope went taut. Each pot seemed to hesitate at the surface before sinking quickly out of sight as the concrete and preceding pots dragged it under. This part of the journey was always a little frightening to me as there was a real danger of the rope snagging an ankle and causing serious injury or pitching one of us over the side. This was one of the reasons I was made to sit at the front of the boat, which was always a little frustrating, at least until I had some animals to examine. No one wore life jackets, and neither my uncle nor my grandfather could swim. In fact, most people in the community were terrified of the water, and even today my mother has nightmares of the sea rushing up to the windows of my grandparents' cottage, over-whelming all inside. I stared into the waters passing by, catching sight now and again of dark beds of wafting kelp interspersed with brighter patches of sand, which appeared pale blue or green from the surface. Below us was another world, which I strained to glimpse through a depth of water that rapidly absorbed the light and cast a veil over the seabed. The pots were ejected in a matter of minutes, and my grandfather, satisfied with where they had been dropped, moved on to the next tow.

As the morning progressed more lobsters were pulled from the depths, including some large specimens. Because lobsters and crabs have a hard exoskeleton or shell they can only grow by casting this off, pumping themselves with water to inflate their body and then secreting a new skeleton. When the new skeleton is first secreted it is soft, and because the animals are vulnerable at this time they remain hidden. The exoskeleton hardens over a few days and then in the subsequent months the lobster or crab will feed, gaining muscle and filling out the new armour. In theory lobsters have indeterminate growth. In other words, they can continue to grow endlessly, although the time between moults extends as they grow older. The largest specimens captured have been over a metre in length and have weighed over 9 kilograms. So somewhere in the depths might be an even bigger monster, Claws instead of Jaws.

As the pots of subsequent tows were pulled on deck, a fantastic variety of other creatures were captured and some thrown up to the front of the boat to satisfy my curiosity. This was my first contact with the alien world beneath the waves, and the animals were entrancing being so different from anything on land. The most common were edible crabs, the classic large orange-brown pie-crust crab that most Europeans are familiar with. My uncle or grandfather would spit in their face, which caused them to curl up their legs and claws, making it easier to handle them. The claws were then twisted off with a crunch and tossed into a bucket, and with the largest pieces of meat now taken, the crab would be thrown back into the sea or smashed and used for bait. If a crab was particularly difficult to pull out of the pot or it managed to nip my uncle, it would be bounced off the gunwale of the boat with a curse and I would watch it spiral down into the depths with a broken shell and a trail of innards left drifting behind as it sank in the water. The treatment of the animals was brutal and at times made me cringe. They were such robust, tenacious creatures, built like tanks, grabbing hold of the mesh on the lobster pots with their massive black-tipped claws to stop themselves from being pulled out. That an animal giving the appearance of so much strength could meet such a careless end felt unfair to me and somehow wasteful. One of my favourite science fiction writers of late has reimagined them as a vicious alien race, giant crabs that eat their young and enslave humans by coring their brains – a terrifying form of crustacean vengeance.

Much stranger animals would also come aboard, and some of these really were straight out of science fiction. Large starfish often emerged with stomachs actually protruding through their mouths, slavering over the bait in creels. An animal that has radial symmetry, where if you divided the creature across the centre in various planes the two halves would be near identical, is not something you see on land. The starfish were the common orange-red variety found occasionally on seashores at low tide or large, spiny grey animals covered in conical spines on top with rows of small, active, pale-yellow tube feet terminating in suckers underneath. If placed on the deck they would rapidly stick to it and crawl slowly into the shade. The tube feet would pull them along like hundreds of little suction cups. Later, during my studies at university, I was fascinated to learn that despite their slug-gish movement these animals are predators as well as scavengers, attacking molluscs, such as clams, scallops and mussels. They envelop them in their arms, prising the shell slightly open with hundreds of tube feet and evert their stomach through the mouth, which is located underneath at the centre of the star, and digest the unfortunate victim externally. Perhaps the most notorious is the crown of thorns, a spiny sea star that eats reef-forming corals and can occur in their thousands, forming a plague. Many years later, during my undergraduate degree, I would sit and listen with fascination in the dimly lit lecture theatre as our head professor, Trevor Norton, described how these animals were collected and cut up in an attempt to rid the reef of these foul pests. Unfortunately, many of the pieces of starfish thrown back into the sea grew back into new individuals. The worst science fiction movie horrors have nothing on sea stars.

We also caught sharks in the creels which, as you can imagine, was very exciting for a young boy who had only ever seen them in films or in books where they were eating sailors cast into the water because of the torpedoing of their ship, or menacing divers on a tropical reef. These were not the great man-eaters so loved by film directors, though. They were small, blunt-headed with pale-grey sand-paper skin on their backs covered with black spots and white beneath: dogfish. Their eyes were always a little disconcerting to me as they were pure black, giving them a soulless look as though they were evil. They were solid muscle and had a disconcerting habit of twisting around if grasped by the tail to bring their mouth towards your hand, which was slightly terrifying as, despite their size, they still had those flesh-cutting teeth. Usually, these were just tossed into a bucket or swung by the tail to have their heads dashed against the boards of the boat, trailing crimson blood from the gills and mouth. They were used for bait, which, as was the case with the crabs, felt like an unfitting end for such a powerful beast.

Towards the end of the morning, by which time I had accumulated an exciting array of animals from the submarine depths below the boat, we discovered a tow with no lobsters and the bait also missing. My grandfather and uncle immediately began to speculate that the pots had been visited by their nemesis, the conger eel. These powerful fish were thought to 'walk the tow', moving from one pot to the other, stripping the bait and ending any chance of catching lobsters. Sure enough, in the penultimate pot, one of these black serpentine animals was brought on board thrashing almost too fast for the human eye to keep pace with. My uncle simply opened the pot and turned it upside down to release the animal onto the deck. They were hated not only for their depredation, but also because they were dangerous: they could deliver a serious bite. The sight of one of the eels was enough to bring a torrent of vicious language from the adults. I was also terrified of this thick, muscled and powerful animal and withdrew to the front of the boat holding my legs up off the floor for fear of being attacked. The unfortunate conger was stamped on and then subject to multiple stabbings through the head with a filleting knife. Somehow it was still alive hours later, the deck of the boat washed with its blood before being tipped back in the sea, unravelling itself and sliding beneath the waves.

Many years later, I would see one of these sleek black fish weave across a seabed of cobbles at a depth of 49 metres diving off the south-western tip of the Isle of Man. It was a rare dive in that we could see the surface from such a great depth and everything was bathed in a gentle golden light like dusk. The eel was graceful and near 2 metres in length, its large black eye staring at us intently as it swam past. It was a powerful moment for me, seeing this animal fully revealed in its own environment and so beautiful.

Once all the pots were hauled and redeployed, my grandfather took us fishing. Now there was the excitement of the hunt. My father and I unwound the two hand lines over the side of the boat and we trolled up and down parallel to the rocks or crossing the bay. My father particularly enjoyed this and we sat on either side of the boat staring in concentration at the sea behind us while my grandfather stood at the engine puffing on his pipe, looking ahead and occasionally grabbing my line to check if there was something there. The lines were baited with an amber eel, a piece of rubber tubing with a hook and a set of brightly coloured feathers with a 'spinner', a triangular piece of metal that spun in the water creating flashes resembling a swimming fish as it reflected the sunlight. The lines thrummed and strained with the movements of the sea and the boat, and sometimes they skimmed above the kelp feeling like small tugs as the hooks ripped through the blades of the seaweed. It was not uncommon for the line to be stopped dead, cutting into my hand and then suddenly releasing. This was a sure sign the hook or hooks had caught on the seabed or on the holdfast of a large kelp, usually resulting in the loss of the bait and lead weight. When this happened the bait was replaced by my father or uncle with supplies we had brought with us for the day from the fishing tackle shop in Sligo town.

When a fish did take the line, however, there was no mistaking it. There immediately came a rapid stutter of strong jerks, and

I shouted to everyone as I struggled to start to pull the line in. Sometimes the catch was lost, but often I would bring in a flapping, leaping fish which would be immediately unhooked by my father or grandfather and dashed on the deck to prevent it propelling itself overboard again before being dropped in a bucket. It was incredibly exciting, the level of hysteria increasing with the number and size of the fish. In those days the ocean seemed bountiful. Then we saw the water churning and splashing, almost boiling, as a large shoal of mackerel pursued baitfish in towards the surface and against the shore. We hooked them six or seven at a time, sometimes the line snaring them by the tail they were so thick in the water. They came over the side of the boat like a mad, jerking puppet show, beautiful fish with pure white bellies and a stunning emerald green to blue back with black chevrons all the way down. They were spindle shaped, streamlined for speed with finely serrated jaws and a large eye with a silver sclera and large black pupil. On one occasion, returning from a similar fishing trip, we had stopped at a small bay to watch several old men and women filling buckets with mackerel off the beach in their hundreds. The waves looked like mercury washing up the shore, solid with mackerel where they had chased their prey right into the shallows and up the shingle. Mackerel are migratory fish and spend their whole time swimming in the upper reaches of the ocean. They spawn at the edge of the continental shelf in spring and early summer, then come inshore in large numbers to feed, which was when we caught them.

As a child, the real prize for me on my grandfather's boat was the pollack, a type of cod. I would beg for my grandfather to keep the boat out if we had not caught one of these superb fish on our trip. Pollack could be two feet or more long, were powerfully built, shaped like an elongate diamond, shiny black along the back fading to a coppery bronze along the sides and belly. They had triangular fins, three along the back and two along the belly, and a broad, dark tail fin. We were on the verge of heading back for the dock when the telltale hit on the line came, stopping it dead. As I hauled the fish in using all my strength, it fought valiantly on the line giving mighty jerks, diving and at the last minute darting under the boat or leaping into the air as I brought it up. They often escaped from the hook, but on this day I managed, with help, to land a fairly large one and everyone on the boat was delighted. Many years later, while diving off Plymouth, I sat in a gully of silver sand surrounded by rocky ridges covered in kelp looking up towards the surface. I spotted two pollack, unmistakable in their outline, cruising menacingly over the rocks and then suddenly lancing down to catch sand eels desperately fleeing for cover. It was mesmerizing finally seeing these animals underwater in their own environment, and also a thrill to finally understand what effective predators they are.

Once our fishing for the day was over, my father, grandfather and I walked back to my grandparents' house with buckets of lobsters, fish and crab claws. We were all cold, damp and tired. Even though it was a day in late August, the wind, the cool sea spray and the motion of the boat drained all of us. The house was always slightly damp and chilly so everyone made for the kitchen which was kept warm by the Aga. Immediately, my grandmother sprang into action, mugs of tea were produced along with soda bread, still warm from the oven, and butter, which melted into it. The fish were immediately gutted, coated in flour and dropped into the sizzling frying pan. The lobsters were thrust live into a large pan of boiling water. They would flick their tails once or twice and my grandmother held down the lid until they were still and rapidly turning from blue to bright red. The kitchen buzzed with excited conversation about what we had caught, what I had seen and whatever gossip was passing about the local community that day. The smells of frying fish, freshly baked soda bread and wet dog emanating from the border collie underneath the table all contributed to the warm glow.

The thrill of hunting in the sea has never left me, although now, as with any biologist or natural historian, it is the hunt for new habitats or new species that excites me. It demands patience and some knowledge of where the life we seek might dwell in the deep. But it is the same excitement, whether it is opening a midwater net to see it jumping with bioluminescent fish and squid or the thrill in seeing an animal from a submersible for the very first time, one you knew was out there in the ocean and which was on your mental wish list of things to see.

Those summer holidays spent on the coast of Sligo and in my grandfather's boat were truly formative in allowing me to draw a path for myself. During those summers by the Atlantic, I learned so much and saw so many strange and incredible species up close in a way many people never get the chance to do. When not in the boat, sometimes we would join my grandmother in some form of gleaning activity on the shore. We would gather sea rods, which are the stalks (technically 'stipes') of kelp, large seaweed that generally live below the level of the tides. They have a root-like holdfast, which attaches them to rock, a long flexible and rubbery stipe, and then a large leafy blade at the top. The stipes were four feet or more long and were washed up on the shore, especially after storms. They would be piled up at the top of the shore for collection as they were used in the manufacturing of alginates, gelling agents used in a wide variety of applications, such as an alternative for gelatine. Sometimes we would even collect seaweed for food. Carragheen or Irish moss is a low-growing red to yellow-brown weed with a flat shape dividing dichotomously to form stubby branches. My grandmother would boil this weed in milk and sugar to make a blancmange, which was set in the fridge. I loved this sweet gelatinous milky pudding. Then there was dillisk, known in England as dulse, a seaweed forming a deep purplish-red series of straps growing from a single point, usually on the stipe of a kelp or on rocks low down on the shore. Dillisk was collected and then dried on the hedges at the front of the house or on the walls. It was then eaten as a snack, like vegetable crisps, mainly by my uncles, who would visit and hoover it up, claiming it had remarkable health benefits.

Armed with books on the natural world, frequently given to me at Christmas by my parents who nourished my fascination, I pushed further and further away from the shore below the cottage and along the coast, rounding the point to the much taller ledges that would slope into cliffs. When I think about what I used to do, my hair stands on end. It is unlikely I would allow my own children to vanish off for the day exploring the rocks with the Atlantic surging in and out of sight of any adult. The smooth limestone surfaces of the horizontal ledges would often remain wet and were covered in an invisible coating of microbial slime. An unwary step sent me flying several times, cracking the back of my head on the flags and resulting in a lump on the head or a 'bourquine' – the latter being the local term for a limpet. There were also chasms where the water ripped up the vertical faces with a sound like tearing metal, and the ocean swells seemed to periodically swallow the lower rocks. I had to descend small cliffs through natural chutes and down steps formed by the limestone, looking from above into huge rock pools formed by water collecting in the corners of the sloping ledges against the next step of rock strata. At the furthest point I could reach, the sea had eaten a great gouge in the land and roared into a huge cave called Culbrain. I would watch the waves here roll in with a deep rumble, shaking the rock under my feet, and the enormous golden-brown kelps wash to and fro on the bottom of this black cavern. On a sunny, gusty day, this was a truly spectacular place.

The ledges here became my favourite haunt. The rock pools contained animals I did not see elsewhere. Purple, black and dark-green sea urchins nestled into depressions they had made in the limestone. They were almost impossible to prise out of their shallow burrows. There were sometimes large chestnut-brown or dark-green sea slugs called sea hares. This name may refer to the large pair of rolled tentacles at the front of these animals and the smaller pair a bit further back on the body. With imagination you could see the tentacles as ears and the head as vaguely rabbit-shaped. My favourite pastime was to lie on the rock holding my fingers in the water. Glassy shrimps, like miniature transparent lobsters, would come out from hiding places and tentatively approach my hand. They would float towards me, propelled by the swimmerets underneath, but, if I moved, a flick of the tail would thrust them away. If I remained absolutely still they would stealthily approach my fingers and begin to pick and pull at the soft skin under my nails and around the quick. I was fascinated by the contact of these tiny delicate animals, the slight tickling sensation of their attentions and the warmth of the sun lulling me into a trance-like state. Occasionally the blennies would come out and try this as well, although they would definitely nip, causing me to snatch my hand out of the pool and breaking my mindfulness.

As I spent more and more time on the strand and out along the ledges, a strange thing happened. I began to regard the treacherous rocks, especially those out of sight of the cottage around Culbrain, as my own, a secret world that belonged to me and me alone. I was completely comfortable with being out in this wild place with no other humans to disturb me. I got to know where all the animals and seaweeds were found, returning year on year to the same pools to find the same shrimps, fish or snails. Sometimes I'd find them changed by a winter storm or a sudden influx of dead kelp. By extension I began to feel the same way about the ocean. I felt, and still feel, that it is mine. Somehow, despite its size and power, it belongs to me. I came to see the ocean as something that other people had no right to damage or destroy.

One day, I came upon my urchin pool and discovered that someone had been down with a spade and dug out the urchins because they had discovered they could be sold. The pool was never the same again, the scars of the spade through the encrusting algae, and broken pieces of urchin lying all around, evidence of the crime. I was filled with anger! How dare they destroy my pool! This is an anger that always boils up when confronted with deliberate or careless destruction of the ocean by people or industry, and it has never left me. Often those who exploit the ocean never seem to realize that they rely on its productivity to maintain their livelihoods.

At 17, intent on becoming a marine biologist, I went to the University of Liverpool to study for an honours degree in Marine Biology. I chose this degree because it entailed a year at the university's marine station at Port Erin on the Isle of Man. I enjoyed it from the outset, learning about the full range of animals in lectures given by Professor Ronald Pearson, a man who interspersed his talks with stories about his and Mrs Pearson's travels and misadventures. We learned about ecology and behaviour, the latter from a lively professor who constantly apologized for sordid examples of sexual selection in dung flies, and in animal physiology classes we learned about how animals adapted to their environment.

The highlights were the field courses, which, in the late 1980s, formed a substantial part of the course. We did two courses on marine ecology and one on algal ecology at the Port Erin Marine Laboratory. I was in my element, as here we learned about the ecology of rocky and sandy shores by going out and surveying them. I remember a crowd of us would-be marine biologists trying to keep up with Richard Hartnoll, a blond lecturer with round spectacles striding across the slippery rocks in his wellies and a long black plastic trench coat, looking more like an East German secret policeman than a scientist. These visits to the shore always seemed incredibly early in the morning when it was cold and grey and many of us were nursing hangovers from the lively evening activities in the pubs or hotel bars in Port Erin. I knew what many of the animals and seaweeds were from the shores of the west coast of Ireland and for me the fieldwork was just a delight.

In the summer of my second year I went to Fort Bovisand in Plymouth, a Napoleonic fortress turned over to the training of both leisure and commercial divers. I did my British Sub-Aqua Club (BSAC) diving courses, bringing me up to Sport Diver level, and the Professional Association of Diving Instructors (PADI) Advanced Open Water qualification. I trained from the outset in a dry suit, made from 8 millimetre neoprene and sealed at the cuffs and neck to keep out the cold water of the English Channel. The two weeks involved various drills aimed at getting us familiar with diving equipment and to develop our skills as divers. This involved learning how to control our buoyancy so it was neutral while underwater, a difficult skill that involved balancing the amount of air in our suits and inflatable jackets so we were neither too light nor too heavy. One meant an uncontrolled journey towards the surface, while the other left a diver struggling on the seabed like a beetle flipped on its back on land. Navigating underwater was always difficult as identifying memorable landmarks in a world festooned with seaweed, where visibility was generally less than 5 metres, is challenging. The secret to finding your way around underwater was to use a compass, something I always found somehow counter-intuitive, and was surprised when it actually worked. Other essential skills included rescue and first aid for divers, where the grim realities of what might happen if a diver ascends too quickly (the bends) or held their breath while ascending (burst lung) were explained to us. It was hard work, and I remember during one rescue drill having to tow a large fireman called Mick a long distance to the wall of Fort Bovisand harbour and physically drag him up the steps. Unfortunately, I did this holding him feet first, only realizing after several groans from my victim that his head was bouncing off the stonework.

It was here, though, that I first got to see marine life underwater first-hand. Seeing these sometimes bizarre animals convinced me even further that marine biology was the career for me. There were brightly coloured sea slugs crawling around on the encrusting life on the harbour walls, wrasse or pollack swimming by or darting off into the turbid waters of Plymouth Sound, but most distracting were the cuttlefish. These relatives of the octopus have a distinct head with a large pair of eyes and thick tentacles hanging down, behind which is a boat-shaped body surrounded by a continuous fin. They would hover in front of us changing colour and skin texture from a smooth black-and-white zebra pattern to a mottled spiky skin. They would sometimes hold their two central tentacles up in front of themselves as a threat display. On a night dive, I found one asleep in a shallow pit on the sand and lifted it up, before it awoke and shot off using a jet of water squeezed through a funnel beneath its head. These animals are voracious predators of crabs and will keep absolutely still as one approaches before shooting out a long pair of catching tentacles and pulling the crustacean in, wrapping it in the rest of the tentacles and delivering a poisonous bite. The paralysed crab is rapidly dismembered and eaten.

We dived the wreck of the *James Egan Lane*, an American Liberty ship that was torpedoed late in World War II and sank while making for the shore, fortunately without loss of life. The bows of the ship were still intact and loomed out of the murk as we descended, free-falling like parachutists into the depths. The feeling of weightlessness was intoxicating, and I felt somehow larger, like a great marine predator myself. The rusting wreck was home to forests of hydroids – creatures like mini white Christmas trees – and large feathered white and orange sea anemones, much bigger than the human hand. There were ling – fish that look like elongated cod – conger eels, usually peering out from a hole or a pipe, and the occasional tompot blenny, small, mottled golden-brown fish with big heads and feathery protrusions coming out of the top. We swam down the hull of the ship to the sand upon which it sat, looking up to see the bows towering above us. Here I was on the set of my own Jacques Cousteau movie. Swimming along we could enter the structure of the ship itself, with the bulkheads of the deck still intact over our heads and a jumble of wreckage and cargo below. The wreck was open, and therefore safe, but the sense of adventure in exploring the gloomy remains held a frisson of danger.

We also did our first deep dive in Plymouth Sound. This was on a chained buoy in Mashfords Bay up towards the naval dockyards. I remember sitting at the bottom of the chain in pitch-blackness, my head humming with nitrogen narcosis. We had been warned about this effect during our lectures on the diving course. Nitrogen has a narcotic effect on the brain when breathed at high pressures. It causes a range of physiological symptoms depending on an individual's physiology and the depth to which a diver swims. It certainly affects judgement and the ability to do simple arithmetic, but in more severe cases a person might feel euphoric, drunk or disorientated. I was certainly feeling 'fuzzy headed', alternately watching a large red scorpion fish and Mick, upside down thrashing his fins. He was more severely 'narced' than I was, and the instructor had to gently turn him the right way up so that we returned to the surface, the symptoms of narcosis dissipating quickly as we ascended.

At the end of the third year of my undergraduate degree, I and a fellow student were recruited to go on an expedition with one of our lecturers, Professor Steve Hawkins, to the Azores to explore the marine life around the island of Faial. We were welcomed by Ricardo Santos, the head of the marine laboratory, and his team of scientists and divers. I suddenly found myself diving in the gin-clear warm waters of the mid-Atlantic on sheer cliffs or in the crater of an extinct volcano. Across a channel from Faial, the huge volcanic peak of the island of Pico a few miles away dominated the view. To me it was an exotic and breathtaking location both underwater and above. Here there were new fish and invertebrates to see and learn about: frightening slate-blue morays mottled with lemon yellow, their jaws gaping open in a threatening display; carpets of low-lying seaweed turf; deep-purple urchins with needle-sharp spines more than six inches long, waving them around menacingly. Sometimes we were surrounded by shoals of large silver fish, jacks or barracuda, or we encountered enormous stingrays, dark grey and as large as a dining table. In deeper waters we would find large bushes of black corals which looked almost silver white, like trees from an elven forest, surrounded by rainbow-coloured fish called basslets with flamboyant fins. Nitrogen narcosis is a significant problem at depths below 40 metres, and we dived to more than 50 metres as the diving regulations of the time were less precautionary than they are now. My bubbles sounded like the high notes on a xylophone as high-pressure nitrogen interfered with my nervous system. The feeling of euphoria, Jacques Cousteau's 'rapture of the deep', was addictive. For someone who had grown up in a commuter town to the east of London, I found myself in a paradise, subtropical islands with sunlit waters teeming with life. This was definitely what I wanted to do for a living.

We dived in sea caves where the walls were painted bright scarlet and yellow with encrusting sponges and we had to 'leap' from behind one boulder to the next because the force of the Atlantic swell coming into the restricted space was so strong. On one dive outside one of the larger caves I received a very painful sting when I accidentally put my hand into the middle of a retracted sea anemone. I didn't know that these animals could sting, and there was a possibility that this animal was new. I flipped it off the rock with my slate, a sheet of plastic used for scribbling notes on with a pencil, and caught the animal which looked like a purple-pink blob covered in small pink cauliflowers. I wrote on the slate that if I was found unconscious the culprit was in my sample bag and headed straight for the surface. My hand came up in white lumps, but the pain only lasted a day or so. An eccentric biologist from the Natural History Museum in London put the prized specimen in a tank and identified it as *Alicia mirabilis*, a nocturnally hunting sea anemone.

As I approached the end of my undergraduate degree, John Thorpe, a genetics lecturer at the marine laboratory who seemed only to be around after midnight and had a fascination with old clocks and antiques, told me he might have a PhD available on nemerteans. I was aware of these animals as I had encountered them under rocks lying at the bottom of the sandy shore in Ireland. Turning one of these over might reveal a heaving, glistening ball of worm, black with a purplish iridescence. They immediately struck me as odd, different from any other animal I had seen before and were therefore interesting. I had purchased a book on British nemerteans on a visit to the Natural History Museum in London by a chap called Ray Gibson. My Irish worms were *Lineus longissimus*, the bootlace worm, which was reputed to grow up to 50 metres in length, making it one of the longest animals on Earth. It turned out that Ray was a co-supervisor of this project, so I was immediately intrigued. I flew over to Liverpool to meet him, a very well-spoken man who went to work in a blazer, shirt and tie but lived all day on cigarettes and black coffee. He was also full of energy and totally obsessed with studying this relatively unknown group of marine animals. He sat in his office surrounded by piles of his papers on them, some beautifully illustrated with paintings, talking about his travels collecting these animals. I was entranced and immediately signed up.

# The Deep

Within months of starting, I began to discover new species of these animals on the shores of North Wales, north-west England and Scotland. It dawned on me just how poorly understood the ocean was. Although a trip to a meadow would reveal a huge variety of insects and plants, the very fact that meadow could be visited by any person day in and day out, as the English countryside had been by naturalists for hundreds of years, meant the biodiversity was well described. That the coastline of Britain, probably the most studied and explored seas globally, could turn up new species based on my searches over just a few months was startling. It also meant that the seas of the tropics or the deep ocean must harbour many, many species never described by scientists and many never even seen by humans. For me, someone who particularly loved unusual marine creatures, this was an amazing prospect. Recently it has been estimated that we have only described 10% of more than two million marine species estimated to be in the ocean.

During my PhD I found myself on the shores of the North Island, New Zealand, searching for nemerteans but discovering a whole new biota I had never seen before with vivid purple crabs and enormous snails. I was a long way from where I'd started on my grandfather's boat in Sligo, but I'd reached the goal I'd set out to achieve: I was finally a marine biologist. The adventure had begun, but there was clearly much work to do. During my degree I learned about pollution of the ocean, which at that time focused on oil spills. In the Azores I heard about seamounts for the first time and how Portuguese scientists were working on the management of fisheries around the islands to try and ensure their sustainability. There was, however, the constant threat of foreign vessels coming into the waters to fish illegally, using destructive fishing gear, and the ever-present problem of overfishing. Outside Portuguese waters were the high seas where it seemed anything and everything was fair game; the Wild West in terms of fishing. I also heard about a seamount not far from Faial with hot springs on the summit which was at diveable depths. I began to learn about deep-sea hydrothermal vents.

Since then, in-between a marriage and the birth of my twin daughters, I have spent almost 30 years obsessively working in the deep ocean – and in that time, I have come to view it as my own. I have always felt that the ocean not only needs protecting, but it is also crying out for exploration to push back the void of ignorance. Ignorance about what lives in the ocean prevents us from managing it sustainably, but, more to the point, our failure to understand it leads to a lack of appreciation of its value. Only with this understanding will humanity come to treasure it.

# 2

# Deep-Sea Hot Springs:

## *The Search for New Species and the Origins of Life*

The ocean holds many secrets waiting to be discovered that have the potential to fundamentally change our understanding of life on Earth and beyond. My own quest to explore deep-sea hot springs – known as hydrothermal vents – in Antarctica in the late 2000s showed me how a single expedition can change our understanding of the ocean, and also how difficult such exploration could be.

This adventure began on a cold grey morning in southern Chile in January 2009. As when heading off on a small boat in Ireland as a child, there was a tremendous sense of excitement as the Royal Research Ship (RRS) *James Clark Ross* prepared to head south. We gathered on the monkey island, the roof of the bridge, watching the crew casting off ropes and making ready on the decks as the ship slowly slipped away from the docks. This wasn't the first time I had sailed on this ship in the Antarctic, though last time it had been from the Falkland Islands. This time we were leaving from Punta Arenas in southern Chile, one of the main jump-off points for Antarctic exploration and cruise liners. All of the gathered scientists were anxious about the coming mission. This was because, according to the terms of the funding we had received, finding the vents was critical to ensure that we would be able to return to the Antarctic on a second expedition to survey and sample the life around them. We were all well aware of what the Southern Ocean could throw at us in terms of weather, which wouldn't compromise the safety of the ship, but would potentially prevent us from carrying out our work. Many of us hadn't sailed together previously, although there were some old shipmates among the scientists and crew who I knew from previous expeditions. The hum of the ship was a sharp contrast to the previous day's stillness in port, and black smoke pulsed from the funnel as the engines started to vibrate through the plates of the decking. Paul Tyler, a tall, bearded professor from the University of Southampton, and a good friend and colleague, was clearly delighted to be underway as he had been planning this mission for several years. I had known Paul since my first postdoctoral appointment at the Marine Biological Association (MBA) in Plymouth, and later my wife, Candida, and I had lived close to him and his wife, Mandy, on the edge of the New Forest, so they had become good friends as well as mentors. Paul's presence on a ship always made for a happy trip as he was eternally cheerful, liked by everyone and was capable of resolving the most difficult of personal clashes or disagreements with diplomacy and tact.

On board were scientists from a variety of disciplines, including biologists, chemists, oceanographers and geologists. They hailed from several different academic institutions in the United Kingdom, including London Zoo, where I was working at the time, the University of Southampton, the National Oceanography Centre and British Antarctic Survey. Our mission was to locate the first deep-sea hydrothermal vents in the Southern Ocean. This had never been attempted previously as the conditions, which included the presence of icebergs and some of the roughest seas in the world, were thought to make it impossible to use deep-submergence technologies, such as robots or remotely operated vehicles. The waters of the roaring 40s, furious 50s and the screaming 60s – the latitudes beyond 40°S – were dominated by extreme westerly winds and had a savage reputation.

Hydrothermal vents are the equivalent of volcanic hot springs on land. They occur in places where seawater seeps into the seabed and comes into contact with hot rock, usually associated with a magma chamber lying in the Earth's crust. The water is superheated, becomes buoyant and rushes up to the seabed, where it exits through chimneys as focused plumes of fluids, appearing like black smoke, at temperatures of up to 400°C, giving rise to the term 'black smokers'. In some cases, these fluids may become mixed with more seawater under the seabed and leak from it as a much cooler diffuse flow. What makes hydrothermal vent ecosystems particularly unusual is that bacteria and other microbes living around the vents feed off chemicals in the hydrothermal fluids, particularly hydrogen sulphide and methane, picked up from the hot rocks lying below. The chemicals supply the bacteria with energy to grow in a process called chemosynthesis. In most other ecosystems this energy comes from sunlight through photosynthesis in plants, algae or bacteria. Animals living around vents either eat the bacteria or host them internally or on external body parts where the bacteria supply them with food in a symbiotic relationship.

Imagine the complete astonishment that the original discovery of hydrothermal vents caused back in 1977. As with many great scientific discoveries it was in large part down to luck. As far back as the 1960s, scientists had predicted that there may be volcanic warm-water springs located on mid-ocean ridges associated with seafloor volcanism, just as on land, where hot springs, geysers and bubbling sulphidic muds are found in areas that are volcanically active. The mid-ocean ridges run around the Earth beneath the ocean like the seams of a cricket ball, and are where magma wells up from beneath the crust forming new seabed. Depending on the underlying geology of the ridge, they may be gently sloped or much more rugged and steep, comprised of numerous volcanoes. In February 1977 an expedition using the American submersible *Alvin* was mounted to search for hydrothermalism in an area of mid-ocean ridge to the north-east of the Galapagos Islands following the discovery of temperature anomalies on the seafloor. The first thing the geologists found among the basalts – volcanic rocks that form the seabed – were bizarre animals, including giant white clams, brown mussels, ghostly white crabs, squat lobsters, giant tube worms and what looked like fuzzy orange golf balls attached to the rock by a thread, nicknamed 'dandelions'. The rocks themselves were discoloured red, orange and ochre, indicative of hydrogen sulphide and exactly what might be expected if hydrothermal circulation was active. Water samples stank of rotten eggs, another telltale sign of hydrogen sulphide.

The occurrence of such large and strange animals was simply not expected and in fact was unprecedented for the deep sea, where lack of food normally limits the size and numbers of animals. On board the US research vessel *Knorr*, the mothership of the submersible *Alvin*, there was a single small jar of formaldehyde which was used to preserve animal tissue for biological studies. There were no dedicated sample jars or other containers for specimens. The clams, crabs, limpets, tube worms and other creatures like flabby eel-like fish were deposited in Tupperware boxes meant for geological specimens or wrapped in cling film from the ship's galley. So short were the scientists of preservation fluids that some of the specimens were pickled in vodka for the journey back to land. After painstaking scientific research and several return trips to the Galapagos vents, it was discovered that the dense communities of animals were reliant on chemosynthetic bacteria for food.

Deep-sea hydrothermal vent ecosystems turned out to be a treasure trove for scientists, and following the initial discovery that chemosynthesis could drive entire ecosystems, more surprises were uncovered. Many of these habitats were what we would regard as extreme, and the microorganisms found living in them were termed extremophiles (extreme-loving). The bacteria around vents can live at temperatures of up to 121°C. Evidence was uncovered that as well as bacteria there was a second major branch of microbial life, the archaea, which occurred in abundance around hydrothermal vents. These microbes showed fundamental differences in their biochemistry from bacteria. In archaea, the cell membranes and cell walls are constructed of different materials from bacteria and the machinery for processing DNA is also distinct, more like that of multicellular organisms, such as humans. These hyperthermophilic (heat-loving) microbes appeared to be the most primitive organisms found to date, falling out at the base of the tree of life. Could it be true that genesis, the origin of life, occurred at hydrothermal vents?

This idea was first put forward in the mid-1980s but was dismissed because it was thought that the temperatures at hydrothermal vents were so high they would destroy the complex molecules that might have given rise to life as soon as they were formed. This all changed in 2001, when the Lost City vent field was found 15 kilometres from the central rift valley of the Mid-Atlantic Ridge, a range of submarine mountains that runs down the middle of the Atlantic Ocean. The vents are low temperature, up to 75°C, and emanate from spectacular bright-white carbonate chimneys, one of which, Poseidon, rises 60 metres from the seafloor. Elsewhere, the vents look like feathery plumes growing out of the seabed, forming delicate towers up to 30 metres high, or grow as fingerlike protrusions appearing like upturned hands.

Scientists realized with excitement that the conditions at Lost City probably bear a close resemblance to those of the Earth in the Hadean eon, a period of time more than four billion years ago that occurred after our planet had formed and cooled down. This was just at the point where the genesis of life is thought to have taken place. The fluids emanating from the Lost City vents were alkaline rather than acidic so their chemistry was entirely different to that of the black smokers. At the alkaline vent system, it was found that energy-generating reactions occur that are analogous to the simplest types of energy metabolism catalysed by enzymes in microbes today. The other important aspect of the vents is that the rock they are made of contains tiny cavities or pores where the products of such reactions could concentrate. Such mineral cavities could have acted like static 'cells', the tiny compartments which make up all multicellular life. However, in order for these 'cells' to escape from their mineral prisons they had to evolve membranes, flexible walls made of fats and protein that enclose the material within, known as the cytoplasm. Membranes are also critical in controlling the environment within the cell by selectively allowing or transporting materials in and out of the cell and allowing cells to communicate with each other. A cell membrane can also invaginate to envelop another cell, allowing predation. That mineral compartments could have been precursors to free-living cellular life is one aspect of the hydrothermal genesis theory that has advantages over other ideas about how life started.

For example, the 'organic soup' hypothesis suggested that the action of lightning in a primitive atmosphere could generate organic molecules. These in turn could give rise, through further reactions, to nucleic acids, which are the material of DNA.

However, the issue with this theory is that there is no obvious mechanism for the chemicals generated in an ancient atmosphere to concentrate and react together and accumulate as the precursors to life. Various ideas were put forward, such as a potential role played by tidal pools or absorption of reaction products onto clay minerals, but none appear as plausible as a role played by mineral 'cells' or the natural chemical and physical gradients formed at low-temperature alkaline vents. Once cells had evolved membranes and had become free living there was no looking back. It is interesting to note that the earliest division of life between bacteria and archaea includes a difference in the fundamental structure of their cell membranes. Perhaps this difference reflects a split in these organisms that reaches all the way back to this ancient time when the first free-living cells escaped from their rocky pores.

During the 1980s to early 2000s, deep-sea hydrothermal vents were discovered in all of the major oceans of the world, down to a depth of nearly 5,000 metres. Over the last 25 to 30 years, new species have been described from these ecosystems at a rate of two per month! The fauna inhabiting the vent fields of the East and West Pacific, Atlantic and Indian Oceans were found to be distinct from each other. In the East Pacific, giant tube worms, clams and mussels crowded the vents. These animals are remarkable for the fact that they rely on chemical energy from bacteria to make their living. The giant tube worm does not even have a mouth or a gut, but takes up hydrogen sulphide, oxygen and carbon dioxide ($CO_2$) through a feathery red plume richly supplied with blood vessels at the top of the worm and transports these chemicals to millions of bacteria contained in a special organ, the trophosome, which produce food for their host from these raw materials. Hydrogen sulphide is deadly to most animals, but giant tube worms have evolved a special haemoglobin, the gas-carrying molecule in our blood that makes it red, that not only transports oxygen and carbon dioxide, but also hydrogen sulphide. So rich is the supply of chemicals around the vents that these worms can grow up to 1.5 metres in a single year and are the fastest-growing invertebrates in the ocean. There are also Pompeii worms, a group of segmented bristly worms that live in tough tubes and perhaps tolerate the hottest waters of all vent animals (up to 60°C). Then there are predators, such as eelpouts (flabby, pallid, eel-like fish) and crabs, the latter of which make a living by nipping pieces off the worms. The West Pacific vents include tube worms and mussels, but also large snails, stalked barnacles and anemones. The snails also have symbiotic bacteria, but these live in the gills rather than a special organ. The Atlantic vents are dominated visually by spectacular billowing swarms of blind shrimps called *Rimicaris,* which cluster around black smokers, but the fauna also include mussels, limpets and other animals. The Indian Ocean vents seem to have a mix of Atlantic-type and Pacific-type animals with heaving masses of shrimps, but also large snails, mussels, dense beds of anemones, and barnacles. There is also a local speciality, the scaly-foot snail, the only animal in the world to secrete its own iron armour. This scale mail covers the red-coloured foot of the animal, protecting it from predators and possibly the harsh conditions around the vents. Something like 70% of these animals, from all regions, are found nowhere outside of vent ecosystems. Adaptation to life in extremes of high temperature and low oxygen, with the presence of hydrogen sulphide and heavy metals – both of which are toxic to most animals – has meant that these species are unable to survive away from vents. Yet scientists found the Indian Ocean vents hosting animals that were both Atlantic and Pacific in origin. So, we were faced with an important question to answer: how do animals living on isolated deep-sea vents disperse and hop from one ocean to another?

The Southern Ocean was one obvious potential route as it connects the Atlantic, Indian and Pacific Oceans and has a vigorous current – the Antarctic Circumpolar Current – flowing from west to east and providing a highway for the eggs and larvae of animals to travel long distances. The key to proving whether or not we were right about this route was to find hydrothermal vents in the Southern Ocean and to discover what kinds of animal life lived around them. This would not only tell us about how life had evolved and spread around hydrothermal vent ecosystems, but it might also provide clues as to how this takes place in the wider deep sea, which is the largest ecosystem on Earth. The enormity of the task ahead of us, alongside our knowledge of the incredible discoveries that we could potentially make on this voyage, meant that the ship was buzzing with a mixture of excitement and anxiety. We wanted to get out there and get the job done.

After a dramatic first few days on the cruise, where we ran over some lost fishing gear floating on the surface of the South Atlantic but thankfully avoided the catastrophe of damaged propellers putting a stop to our explorations altogether, we reached the southern side of South Georgia where we deployed a Conductivity-Temperature-Depth (CTD) sensor on a rosette of Niskin bottles – water sample collection vessels. This was a round frame of titanium scaffolding ringed with large dark-grey plastic bottles with snap-shut lids on the top and bottom that could be triggered to close at the required depth below the ship to collect water samples. Sensors were also mounted on the frame, and the whole array could be lowered over the side of the ship on a steel cable. The CTD was deployed to try to detect signals of the chemical methane, which might be emanating from another type of chemosynthetic ecosystem, a hydrocarbon seep, located near the island. At seeps, hydrocarbons, including methane, leak from the seabed and are also used by bacteria as a source of energy. Thus, they can host rich communities of animals, some of which are similar to those found around vents. Evidence of frozen methane, also known as methane hydrate, an ice-like substance that burns if lit, had been trawled up several years earlier by another research expedition. The CTD returned with water samples that were divided between the chemists for analysis of methane concentrations and to me for filtering for bacteria. The latter was a very long and tedious process of pulling water through small filters which retained particles as small as 0.2 micrometres, small enough to capture the tiniest bacteria. The filters were then frozen at -80°C to be transported back to British Antarctic Survey where a colleague would sequence their DNA so we would be able to identify what bacteria were present. It was the first step in identifying how similar or how different the biodiversity of Southern Ocean seeps and vents were from those elsewhere in the world.

We steamed on south, eager to reach the East Scotia Ridge, which is a craggy submarine feature running from north to south in the Scotia Sea just west of the South Sandwich Islands. It was here, some years before, where the chemical signals of manganese, indicating hydrothermal vents, had been detected in water samples.

On arrival, we deployed the CTD again. Paul, Rob, the Chief Scientist of the expedition, the other scientists and I clustered around the computer monitors showing the traces from the sensors. These appeared as glowing coloured lines crawling across graphs against a dark background. Doug Connolly, a wiry and energetic chemist was in a chair in front of the monitor. The chemistry of hydrothermal vents was his speciality. At 2,000 metres down the readings on the transmissometer – a sensor attached to the CTD to detect particles from hydrothermal vents – and the ESS – another sensor, which measured the oxidation of chemicals in seawater – began to creep up. Doug excitedly pointed out that it was exactly what we might expect to see over a vent. Near the seabed the instruments went crazy, and Paul anxiously called for the winch to be stopped. We began to haul the array back in for worry that it may have hit the rocks on the bottom. When the samples arrived back on board I began the arduous task of filtering seawater again, while the crew prepared to launch BRIDGET (British Mid-Ocean Ridge Initiative Towed Instrument), an array of sensors for detecting vent plumes towed behind the ship. By the time I emerged, Paul and the others were gathered around the computer screen again. I could feel the buzz of excitement in the air. A quick look at the computer screen showed that the coloured traces for temperature and the transmissometer were spiking erratically as BRIDGET passed over the seabed.

Paul broke the silence of concentration: 'We're passing through vent plumes rising from the seabed, I'm sure of it!'

Doug replied: 'They are plumes. They're wandering about a bit, probably with the tide, but they're definitely there.'

He was grinning uncontrollably. It took me a few minutes to absorb the information and its implications before I felt the infectious excitement rise in my chest. This meant we had confirmed that the East Scotia Ridge hosted hydrothermal vents. This was incredibly important as we needed to find the location of the vents to justify getting a second expedition for survey and sampling using a remotely operated vehicle (ROV), a robot that could be deployed from the ship to sample the seabed. Paul and I left the computer room and finished the day with a quiet gin and tonic at the bar to celebrate my birthday, which was the best I could do being as far as I was from Candida and my two children.

Overnight more vent plumes were tracked. These can be pictured almost as plumes of smoke from chimneys rising into the sky, but instead they are plumes of particles of metallic sulphide and other chemicals rising from hot hydrothermal vents on the seabed, which was 2,500 metres below us. BRIDGET was flying through these plumes picking up changes in the particle concentrations and chemicals in the seawater. The problem was that as the currents below changed with the tide, the plumes were also changing direction unpredictably, which made it difficult to pinpoint where the vents actually were. It was like looking for the proverbial needle in a haystack. Vents are generally relatively small, a few hundred square metres, and the ridge and the wider ocean vast. As any scuba diver who has tried looking for something dropped on the seabed knows, it is extremely difficult to find an object underwater where visibility is limited, even if you have a rough idea where it is. The seabed may slope or be broken by gullies, or covered in rocks festooned with marine life. This was the same problem, only we were trying to find the vents by dangling a sensor 2,500 metres down over the side of a ship less than 100 metres long. I sat at the computer monitors logging the position of BRIDGET as we hauled it up and down through the water, between 2,200 and 2,500 metres depth.

During the afternoon, as we approached a deep Southern Ocean depression, the weather began to deteriorate. The increasing height of the waves meant that BRIDGET had to be retrieved, but when it was brought up on deck we found the wire had a cat's paw in it – a kink, in non-marine speak – which was probably caused by the spinning of the instrument as it was towed behind the ship. The cable was frayed and would need cutting and re-termination, a difficult job that involved reconnecting all the electronics in BRIDGET with the core of wires within the sheath of the armoured cable. A CTD drop was managed before the state of the sea became too bad to do any further work and I had to filter the water samples holding onto the apparatus to stop the movement of the ship depositing the glassware on the floor of the laboratory. Doing anything in these conditions is physically draining as you constantly have to brace one way or another as the ship is moved about by the waves. Many times I thanked my childhood spent on my grandfather's smelly boat for my immunity to seasickness.

Southern Ocean storms are notorious. The wind and waves travel around the Antarctic continent from west to east with no land to block them. The fetch is limitless and a depression can build mountainous seas with huge waves called 'grey beards' or 'Cape Horn rollers', legendary among sailors of old. Frank Worsley, the captain of the *Endurance* during the 1914–1916 expedition led by Ernest Shackleton (in which the ship was crushed by ice) described the ability of these huge waves to smash boats 'like eggshells'. Even in modern times, the Southern Ocean represents a formidable obstacle to navigation. In the 1996–1997 Vendée Globe, a single-handed round-the-world yacht race, three of the 60-foot boats were capsized and wrecked by ferocious storms in the Southern Ocean. Derek Lundy, in his book *Godforsaken Sea*, described how some of the waves that smashed down on these yachts were as tall as eight-storey buildings. A Canadian sailor, Gerry Roufs, disappeared during one such storm south of Point Nemo, the position in the ocean that is furthest from any land, at 2,688 kilometres from the nearest coast. He was never seen again, although his yacht was washed up on the coast of Chile six months later.

I had experienced storms before. I had been stuck in the mouth of the English Channel for several days on the RRS *Discovery* in gale-force winds and had come out from the Azores into the teeth of the tail-end of a hurricane on RRS *Charles Darwin*. However, I had never seen such huge seas. I felt both a feeling of awe at the power of such gigantic waves and a sense of helplessness. There was a serious bout of rolling and then a shuddering felt through the deck below my feet, which was a worryingly unnatural motion for the ship. The vessel finally hove to, just sitting with the bows to the wind to ride out the storm.

As the storm raged on, the ship pitched, rolled and juddered through the night. By 4.00 a.m. I was having to grip the edges of my bunk to prevent myself being flung out, the motion of the ship had become so violent. For the first time in my life, I became a little apprehensive at the ferocity of the sea and the protection offered by the thin steel envelope of the ship. As I lay there, all sorts of odd thoughts clouded my exhausted mind. What would I do if the ship foundered? Put on warm clothes, grab the chocolate stashed in my drawer, don the survival suit? Surely if something happened it would be so quick I would stand no chance of escaping from my cabin, down corridors and up the stairs to the lifeboat station. As the ship pitched again more violently than before my stomach filled with butterflies. It's an uncomfortable feeling when a familiar environment suddenly becomes precarious.

Thankfully, several hours later the waves seemed to calm, and I rose and struggled up to the bridge to find out what was going on. The storm had blown us 60 miles off the part of the East Scotia Ridge that we were working on. An announcement was made on the ship's tannoy to hang on as we had to turn around to head back to our previous position. The captain waited for a reasonably flat piece of sea then swung the ship neatly around and we headed for where the vent plumes had been detected.

By the afternoon we were back where we needed to be and BRIDGET was dropped into the water, but outside the weather was deteriorating again and frustration and fatigue began to tell among the team. We had already lost much of the day retracing our steps and trying to sort out issues with equipment, the last thing we needed was another storm that would delay us further. Many people turned in to get some sleep after the battering we had received the previous evening. This left Paul, Rob and I chatting in the main computer control lab until midnight. We were discussing plans for the next few days when suddenly there was the jaw-gritting screech of tearing metal and the whole room started to shudder under our feet. This was immediately followed by every alarm in the winch control cab next to the computer control room going off. I stopped breathing and just stared at Paul, who had also frozen mid-sentence.

'That sounds like a broken winch,' Paul said, the gravity in his voice breaking the spell that the convulsion through the ship had caused. The deck engineer who was controlling the winch called out: 'I've no idea what's just happened. The winch just stopped dead.'

The tension in his voice was clear. He didn't want anyone to think he was the cause of whatever was happening to the ship.

We looked around the corner, lights were flashing all over the control panels. Some of the crew appeared saying they had heard a loud bang from somewhere below, probably the winch room. Outside in the dark it had begun to snow heavily. It was a depressing scene. Word came up that the bearings had gone on the main winch storage drum. This was holding about 7,000 metres of cable, of which 2,000 metres was hanging off the back of the ship with BRIDGET dangling on the other end. Our concern about the impact on the work we were doing was shifting now to worry for the safety of the ship. We knew that if the weather deteriorated further there would be no choice but to clamp off the wire and cut it, dropping BRIDGET and two kilometres of steel cable on to the deep seabed. It was a major blow. With the weather deteriorating and BRIDGET still in the water the situation was serious. Anxious and tired we agreed the only thing to do was to try to get some rest, though I'm sure none of us slept a wink that night. We'd already faced several obstacles on this mission, including entangling some lost fishing gear on the ocean surface and the foul weather, and thankfully overcome them. This one, however, felt fatal to our quest. Perhaps our luck had finally run out. Only the morning would reveal our fate and so another night was spent in restless sleep.

At sunrise we all gathered around the video monitors showing the main winch surrounded by white boiler-suited engineers, peering into spaces with torches and hefting tools about. Bridget was still dangling out on 2 kilometres of wire. However, by the time breakfast was over they had started to wind in the cable. We all held our breath, silently praying that we would recover BRIDGET safely on board. In it came, a few metres at a time at first, and then more rapidly. It was only once we had her back on board that we could all breathe a collective sigh of relief. The ship was safe and we had recovered the equipment that could so easily have been lost. We weren't out of the woods yet, though. The main winch and BRIDGET were out of action for the rest of the expedition.

A key phrase for science at sea is 'adapt and survive'. As we took on the new reality of our situation a discussion in the computer room led to the decision that our only choice was tow-yoing. This involved dipping the CTD and water sampling rosette up and down in the water column as the ship moved forward very slowly so that we might detect vent plumes down below by changes in water temperature and particle concentrations. Doug enthusiastically explained that he had done this before and we should still be able to find the vents. The CTD-water sampling rosette, which was deployed off the side of the ship, was still operational as it ran off a separate winch system to BRIDGET. Several sensors were swapped from BRIDGET to this system and we began the arduous process of dipping the CTD up and down in the water to find the vent plumes.

The following day we deployed our towed camera system, SHRIMP (Seafloor High-Resolution Imaging Platform), in the area we had detected plumes after fixing the winch for this system which was on the deck of the ship. The control box had become filled with water during the storm so needed emptying and drying out. We dropped SHRIMP off the side of the ship and gathered inside around the TV monitors to look at what was below us. I was excited. This would be the first view we would get of the actual seabed on the East Scotia Ridge. Up to then we had simply been retrieving water samples or watching readings from instruments. There was a lot of marine snow in the water, lumps of fluffy organic detritus made up of dead cells of algae, planktonic animals, faecal material and other remains from the ocean surface, sinking to the bottom. This was food for the deep sea. Down at 1,500 metres, we spotted some large squid on the monitors, and as we approached the seabed we saw several very beautiful abyssal jellyfish, one a flattened drum shape, a velvety deep purple-red in colour. Then the seabed appeared. It was covered in pillow lavas. The shapes were amazing, many of the lavas literally froze when they were squeezed out of the seabed and came into contact with the ice-cold water. The result of this was that the hardening surface had cracked and exfoliated, its skin crumbling away. Solidified lava lay everywhere, sometimes looking like piles of enormous striated ropes, hands or even skulls. It was a landscape straight out of the movie *Alien*, when the crew of the *Nostromo* had fatefully landed on an unexplored world to find an alien spacecraft.

Here and there were soft corals, many were quite simple whip-shaped colonies, while others looked like flat Christmas trees. One was an eye-catching deep purple. Rattail fish appeared now and again, typical denizens of the deep seabed with a big head and skinny tail. There were also stalked sea lilies, relatives of the ancient crinoids, growing on the rocks. Small, white brittle stars appeared here and there on the sediments or on the petrified lava. There were also large spiny red brisingid sea stars with their arms extended out into the seawater to trap particles of food. This was the normal 'background' fauna of rocky seabed of the deep Southern Ocean. We started to hunt for vents. First we tracked around the area where we had got the strongest vent plume signals, but did not find anything that even remotely resembled a vent chimney. Usually there is a zone around hydrothermal vents where the seabed fauna is dense because of the higher concentrations of bacteria and organic food. We found no such halo of life, no hydrothermal sediments and no chimneys. It was eerie, and I was unsettled by the lack of hydrothermal activity. Had we really been detecting vent plumes from BRIDGET and the tow-yo or was it something else? Eventually it got to evening and we recovered SHRIMP before dark. Doug was puzzled. The vents should have been in this spot. He decided to spend the night hunting for more vent plumes by tow-yoing with the CTD.

By morning, some good signals of vent plumes had been detected and the water sampling bottles had been retrieved. We ran SHRIMP over the ridge line and then further, crossing several wide and deep fractures in the rock. Paul and I stood with the rest of the team around the monitors in the computer room with our eyes glued to the screen. Just before we got to the position to turn I noticed that we were suddenly getting a high density of the purple soft corals and a lot of small white animals, possibly limpets, the latter on the surface of sediment. We began to make the turn when one of the other scientists from Southampton said: 'Is that bacteria under the rock?'

I looked closer, and sure enough there was a fine, velvety, purplish-white coating under the rocks visible in the oblique camera positioned at the front of SHRIMP. Then we started to see anemones, a typical inhabitant of vents in the Atlantic. Then strange bright yellow-and-white squat lobsters. This had to be it – lobsters just do not occur in Antarctic waters, they are too cold!

# The Deep

I could feel my heart hammering in my chest as the atmosphere switched to one of excitement. The sediments lying around the site were black. We came across a few smallish chimneys, at least one with some diffuse hydrothermal fluid flowing out of it. We had found a site, but where were the high-temperature chimneys? We knew they had to be there because of the signals from the sensors on the water sampling rosette. My brow furrowed as I stared at the screen, as if my eyes could will black smokers into existence. I knew they were there, if only we could find them.

We drifted back over pillow lavas out of the hydrothermal vent area. There were monitors on the bridge too so everyone on the ship could see exactly what we were seeing. All of the scientists and some of the crew had gathered around the monitors, everyone eager to see what would happen next. The suspense of being so close to locating what we had set out to find, but not quite there yet, was almost unbearable. We drifted over a large chasm, probably a fissure, and then back onto the vent field. Mats of bacteria were everywhere on the stones and black sediment. We drifted back onto another part of the site and saw an amazing chimney, like a great mushroom capped in pearly white crystal. More pinpoint adjustments of the ship's position were made and we held our breath. We were so close now, I could feel it.

As the camera manoeuvred into place, seconds felt like years. Everybody was on tenterhooks. And then, just like that, the most wonderful sight filled the screens and our eyes. We were suddenly in a field of amazing hydrothermal vent chimneys, 10–20 metres tall, most capped with white crystal. It looked other-worldly. It was as though some fey hand had crafted a garden of delicate crystalline columns to a design that escaped human comprehension. James, the engineer in charge of SHRIMP, carefully manoeuvred the vehicle through the columns. We saw clusters of the squat lobsters, crowded into crevices and cracks on the rock around the chimneys, as well as hundreds of anemones. It looked as though there was a combination of animals here that had not been seen anywhere else. After a moment of awestruck quiet, peppered with soft, delighted gasps, the room erupted with animated chatter and speculation about what the cameras were revealing. Then, finally, as James steered SHRIMP among the chimneys, compensating for the swell moving the ship up and down, a chimney came into view pumping great clouds of black smoke from two vents on top.

'Yes!' I cried, and joined in the frenzy of shaking of hands and slapping of backs. This was what we had come to the Antarctic to find. No one else had seen a high-temperature hydrothermal vent in the Southern Ocean! It had been known that there were vent plumes emanating from the East Scotia Ridge for 10 years, but no one had attempted to find them. We were the first to set our eyes on such a sight, and I was beyond elated. More folk joined us and the ship became a riot of people shouting and jumping up and down. This was the icing on the cake: we had beaten disaster, storms and broken machinery – everything the Southern Ocean had to throw at us – to finally discover its secrets. This was the key to understanding how life on hydrothermal vents was distributed globally and perhaps how, over timescales of millions of years, animals dispersed from one part of the deep ocean to another. However, the initial glimpses of the vent fauna of the East Scotia Ridge raised more questions than it answered. We were not expecting to see animals like squat lobsters on the vents. They did not exist on Atlantic hydrothermal vents, and animals we were expecting to see, like vent shrimps and mussels, were completely missing. It was like finding the final piece to a jigsaw puzzle, but on putting it in place, discovering the rest of the picture was wrong.

After much applause for James and the ship's crew, we gazed at our vents a little while longer until Paul broke the spell they had cast on all of us by telling us we needed to move on.

'We don't want to damage the chimneys with SHRIMP, we'll be back with *Isis*.'

We all agreed and reluctantly tore our eyes away from the chimneys that had transfixed us. They were made of fragile metal sulphides, and if we weren't careful, SHRIMP could act as a wrecking ball demolishing the delicate structures we were viewing. *Isis* was the UK's remotely operated vehicle, a deep-sea robot tethered to the ship that could dive to 6,500 metres, collect samples and survey the vents with high-definition cameras. It could manoeuvre very precisely in all directions using thrusters so it was the right tool to explore the vents in detail and collect some of the animals we had so tentatively glimpsed. However, we weren't done yet. We needed to find one more vent site on the southern part of the East Scotia Ridge so that our planned second and third cruises to survey and sample these new ecosystems would be secure.

\*    \*    \*

The following day we set out to achieve this goal and reached the area in our next targeted region, where hydrothermal signatures had been detected in an expedition back in 2000. There was a massive pit crater in the ridge, a startling hole nearly 2 kilometres across and hundreds of metres deep that was formed when a lava chamber drained and the roof collapsed. This was nature on a massive scale. Linear fractures stretched away to the north running along in the same direction as the ridge. We began the task of dipping for vents again by tow-yoing and immediately picked up strong signatures of hydrothermal plumes. However, excitement quickly gave way to frustration and concern. The vent plumes were moving around in the current again making it incredibly difficult to pinpoint where they were originating on the seafloor. Hours crept into days watching endless tow-yos with the water samplers and sensors. The weather was unpredictable, and often rough, which added to the exhaustion as people lost more and more sleep.

After two days, SHRIMP was dropped down onto the ridge. The seabed was mainly silt and glassy sand, with intrusions of glassy sheet lava. The glass glittered in the lights of SHRIMP and we saw a lot of animals, including bright pale-purple sea cucumbers, numerous white brittle stars, bright red-and-white brisingid sea stars, bright yellow stalked crinoids, the odd octocoral and snake-tentacled purple-pink anemones. The stalked crinoids – sea lilies – were particularly fascinating as I had seen fossils of these relatives of sea urchins in the limestone flags on the coast of Sligo in rocks 350 million years old. They had died out in shallow waters over time, but here they were in the deep sea, a bright yellow stalk ending in a 'bloom' of branches radiating out from the mouth like the delicate petals of an exotic flower. The animals fed on particles of organic detritus drifting past. The Antarctic fauna in many ways resembled that of ancient seas as it generally lacked shell-breaking predators like lobsters and crabs, allowing communities of particle-feeding animals like crinoids, sponges, corals and brittle stars to thrive. There was also the odd rattail fish hovering above the seabed, snouts pointed downward, the scavengers of the deep sea. We searched around and eventually found the edge of the crater, which I had nicknamed the Devil's Punchbowl, after a sea cave in Sligo with a collapsed roof that opened as a hole in a field near my great aunt's house. There was a sheer cliff dropping 200–300 metres comprised of rubble and small basalt and glass blocks, all coated in a powdering of a bright yellow deposit, some with whitish edges. Under the blocks was a purple sheen that looked much like the bacterial mats we had seen at the previous vents. The bacteria had to be consuming some form of chemical energy, so perhaps the whole cliff face was seeping hydrogen sulphide? The vents, however, were elusive. Icebergs added to the difficulties, drifting past, some with a peppering of distant penguins, forcing the ship to change direction at short notice.

I fell asleep several times watching the monitors and logging the progress of the tow-yos. I had reached the point of exhaustion at which it was possible to find yourself suddenly dreaming while standing up, reality blurring with surreal hallucinations. Adding to my state of exhaustion was worry about my daughter Freya, now at home in Cambridgeshire. The last time I'd been able to call home, Candida had told me she was due to take her to hospital for an important X-ray and a check-up on her recovery from a hip displacement operation. I was desperate for some news on how she was doing, as this was the point where the surgeon would see whether any further operations were required. Her previous surgery had been tough and the pain of those months was still with me. My eyes burnt with lack of sleep and the skin of my face ached and felt slack and rough. Then, after six days hunting for the vents, a dire weather forecast forced us off the site and we went full steam ahead for the South Sandwich Islands to seek refuge there.

The ship was rolling so heavily the following day it made me dizzy. The detour did mean, however, that I could finally get news from home. Freya was healing well. No further surgery was needed, just a close eye on her development over the next few years. I almost cried with relief on hearing the news and left my cabin feeling grateful and uplifted. I had a glass of red wine to celebrate at dinner with Paul who knew what we had been through over the past 18 months. All was well at home and I could return to what we were there to do with a lighter heart. Now we just needed the weather to clear to resume our hunt for the hydrothermal vents at the second site on the East Scotia Ridge.

On the way back from our detour we reached Kemp Seamount, an underwater mountain formed by a volcano off the main South Sandwich Islands and, after taking water samples, we dropped SHRIMP onto what looked like a small volcanic cone on the seamount. It was covered in small white brittle stars, touching arm tip to arm tip, an unworldly sight. There were also larger pale-purple brittle stars and giant yellow sea anemones as well as large black sunstars, bright yellow and orange predatory starfish (at least two species), and various other animals, all living among one another as one big technicoloured community. We were glued to the monitors. Very shortly after starting the survey line, I spotted purple jellyfish trapped on the seabed, and, excitingly, it looked like several had been snared by the anemones. There was the suggestion in scientific papers that seamount food webs could be fuelled by the trapping of animals living in the waters above the seamount. As the dive progressed we saw more and more of the unfortunate jellies, some looking fairly dead, others still pulsating, top down on the seafloor.

We passed areas where there were plough marks on the seabed from grounded icebergs, and other areas where there had been sediment slides. Here and there whip corals were seen, some trembling in the vigorous currents sweeping over the seamount. As we progressed between a saddle lying between two peaks we came across an area where the giant anemones were particularly abundant and here several were in the process of actually eating the jellyfish. One particularly plump specimen was closed, clearly having just ingested the entire jellyfish. This was fantastic! It was proof that the seamount fauna could feed off passing animals in the water column, so the hypothesis of the coupling of the two systems was right in this case. We were all delighted, especially after such a severe pounding by the weather, to be making some scientific observations again. As we passed over the summit into deeper water we came across dense beds of bright yellow brachiopods and then, as animals became less abundant, large, bushy colonies of yellow sea fans with bright-white brittle stars living among the branches. I had written about seamounts before, frequently identifying them as biological hotspots, but this was the first time I had seen just how rich these habitats could be.

After exploring more of the South Sandwich Islands, Paul was anxious that we get back to finding the second hydrothermal vent site. Time was running out, with just two or three days left before we had to steam north, and we needed to find the second vent site to guarantee our return. We decided to head back to the southern part of the East Scotia Ridge where the elusive vent plumes had been detected. At 5.40 the next morning, SHRIMP entered the water and by lunchtime we had come across several fissures. We had found an area of hydrothermal rubble and yellow-orange sediment with several extinct chimneys. The ship was electric with the thrill of what this could mean. Everything was stacking up in favour of us finding what we were looking for, but as we searched further we remained empty-handed. The clock was ticking: we had a deadline of midnight before we needed to move further north to take the first steps towards home. With no sign of an active vent nerves were fraying and spirits were plummeting.

As we got down to the final hours there were just a few of us left at the monitors, the rest having succumbed to exhaustion. It was all or nothing at this point, so I suggested we strike a line south-south-east through an area of inactive, dead chimneys and then on towards the crater, but between the two fissure systems. We upped the speed of the ship to 0.4 knots to cover more ground with the towed camera system. As expected we passed over the dead vent site, but then we hit a patch of rocks with a cluster of seven-armed starfish, animals that none of us recognized as being part of the usual Antarctic deep-sea fauna.

'Hey – I think we're in the far field,' John Copley, one of the scientists from Southampton, shouted suddenly.

We looked intently at the video screen. Black rocks flashed past, powdered in bright yellow, and suddenly we were plunged into rampant life. I was so stunned by what I was seeing that my mind struggled to take it all in. We'd been on the verge of losing all hope that we'd reach our goal, and now we had not only found what we were looking for, but we were seeing one of the most incredible sights imaginable. It looked like a fantastic garden created by the elves in J. R. R. Tolkien's *The Lord of the Rings* or maybe the forest from James Cameron's 3D blockbuster *Avatar*. There were clusters of large anemones, some deep pinkish-red with white columns; animals on stalks formed bunches, some grey-white, others yellow – we realized they were stalked barnacles; and clusters of the squat lobsters familiar from the northern site but in larger numbers. Heated water swirled around some of the aggregations of these crustaceans, not mixing with the cold water surrounding the vent and creating a shimmering almost gelatinous effect in the water. There were smaller creatures down inside crevices: big dense clusters of dark-grey snails. The room went into pandemonium as more and more sea life flashed in vivid colour across the screen. The flow of vent fluid was clearly coming from between the rock strata – it gave everything the look of neatly arranged flower beds. We criss-crossed the site several times finding white, diffuse, smoky fluid emanating from clusters of barnacles, and at one point wisps of black smoke emanating from a big cluster of squat lobsters. We'd found our second vent site and it was even more magical and mesmerizing than the first.

I'll never forget the jubilation that swelled throughout the ship on that occasion. In my exhausted state, as Queen's 'We are the Champions' and 'Firestarter' by the Prodigy rang out from a ghetto blaster in celebration, the euphoria was intoxicating. I stood, swaying at the map table, where Paul joined me grinning from ear to ear. In the mix of the many emotions I was feeling in that moment was the fact that I was very proud that I had played such a major role in finding the vents. Although we had got a brief glimpse of the vent site in the northern part of the East Scotia Ridge, this one clearly showed us that what we were looking at was completely different from what had been seen at vents elsewhere in the world.

Now that we were here, seeing it in all its stupefying glory, I wanted to sample the site right away, and it was very frustrating that we couldn't. We knew that the second and third cruises were now secure, though, and we would be back soon with an ROV. I was intrigued by what we had seen, it was not what we were expecting, and my head was full of the possibilities the views from SHRIMP had given us. Were the Antarctic vents just completely different from everywhere else that had been sampled? Did this do away with our ideas of the Southern Ocean as a highway for dispersal of animals across the other oceans? My mind was buzzing with the possibilities as we feasted on the views SHRIMP gave us of our newly discovered underwater gardens. We went to the bar very excited to celebrate with a drink and to try to calm down. Paul pointed out that we had missed the site on one of the earlier SHRIMP surveys by less than 200 metres. But that didn't matter now, the job was done. I headed back to my cabin and emailed Candida with the news before falling into my bunk – it was 2.15 in the morning and I was asleep before my head even hit the pillow.

I didn't think anything could top my first glimpse of the hydro-thermal vents on the southern part of the East Scotia Ridge, but my hopes were high when I found myself back there a year later with the equipment we needed to explore these wonderful works of nature further. Paul could not be on the ship and had asked me to be the scientific leader of the cruise back out to the Antarctic on the RRS *James Cook* equipped with *Isis*. This deep-sea robot or remotely operated vehicle was about the size of a large SUV with bright red and yellow buoyancy foam atop a metal frame-work stuffed with instrumentation, and two large steel claws at the front along with an array of lights and cameras. It resembled some sort of giant mechanical crab itself.

It was day 20 of the trip and we had launched *Isis* that morning. The atmosphere in the control van of this high-tech vehicle was one of tense expectation. The van, a large transport container, the type you may see on an articulated truck, was blacked out, most of the light coming from two banks of monitors arrayed along one side. The top row of five large monitors showed views from the cameras of *Isis*, revealing dark-blue waters lit by the high-powered lights of the ROV at a depth of more than 2,300 metres. All eyes were on the monitor from one of the forward cameras currently pointed downward as we were nearing the seabed. The lower row of computer monitors displayed the ROV's position relative to the ship, the vehicle's sonar and other technical information, monitoring *Isis*'s vital signs. The faces of the ROV pilots and technicians seated in front of the monitors were lit a ghostly white and the hushed chat related to the final approach of the vehicle to the seabed was barely audible above the sounds of cooling fans from computers and banks of video recording equipment behind and to their right. On this expedition, in January 2010, *Isis* was our eyes, ears and hands on the seabed. Sitting behind the ROV techs were three scientists, eyes glued to the monitors, recording observations on laptops. A number of us were crowded to the left of them for this first view of one of the high-temperature deep-sea hydrothermal vents on the southern part of the East Scotia Ridge. I was tense with the anticipation of what we might find.

'There's the seabed,' said Doug.

We all strained forward to see what the camera revealed. Something materialized slowly from the darkness in the lights of *Isis*. To my eyes it looked like a pile of skulls nestling among the black-and-grey broken basalt of the ridge. The skulls resolved into millions upon millions of lobster-like crabs, each about the size of a fist, all a dead white to pale yellow in colour and jostling for space. These were not squat lobsters, though. They were yeti crabs. There was a collective intake of breath from everyone in the van and then excited laughter. It was the most amazing spectacle. A sight that nature had kept hidden from humankind until that very moment. No one had ever seen this species of crab before our expeditions, let alone in such vast numbers. We had the feeling that we were here as explorers as well as scientists. It is easy to forget that we have only seen a tiny fraction of the deep ocean. What other treasures might be hidden in the abyss? There was mad chattering and gasps as the cameras panned around the mass of crabs, six or seven deep. Nicolai Roterman, my PhD student, was speaking in hushed tones into a recorder for the BBC describing the scenes of both the new vents and the excitement of the scientists as they congratulated each other on what was an awesome discovery.

We manoeuvred the ROV into a good viewing position. The heaps of crabs heaved and rippled in slow motion, probably disturbed by the wash from *Isis*'s thrusters. Warm, shimmering water drifted up from beneath and around the crabs through which pinnacles of glittering basalt emerged here and there from the sea of crustaceans. The yetis had a smooth, white carapace (shell) protecting the dorsal surface while underneath three pairs of stout articulated legs appeared from a dense mat of fine hairs. There was also a very long and robust pair of pincers or claws projecting forward and covered in rounded tubercles and spines, like marine boxing gloves, which the crabs used to fight for space by clubbing and grabbing at each other in an attempt to dislodge opponents from prime spots of vent fluid flow. They were also blind as they spent life in total darkness, but had two pairs of hairy antennae, one long and one short, presumably used to 'smell' and feel their way around like the lobsters we had fished in Ireland.

They were named yeti crabs after the dense hairs that covered the underside of their bodies, giving them a furry appearance.

At the time only one other such crab had been seen on vents in the South Pacific and that one had hairy pincers or claws. Our yeti crabs, with their furry bellies, were a new species and Nicolai immediately dubbed it the 'Hoff crab' after the well-known *Baywatch* actor David Hasselhoff, famed for his hairy chest. Subsequent to the cruise a BBC correspondent picked up on the nickname of 'the Hoff' and a video of it went viral with news channels from across the world announcing the discovery of the first deep-sea hydrothermal vent animals from the Antarctic. David Hasselhoff even tweeted about the new Hoff crab, introducing it to his fans, and it ended up in *Guinness World Records* as the first yeti crab found in Antarctica.

As well as the Hoffs there were also large blue-black and brown snails with red feet, mainly living on vertical rock faces, sometimes hanging down in chains. There were also stalked barnacles in clumps among the crabs or on top of rocks. Hidden in cracks and crevices were bloated-looking pearl-white and pink sea anemones, as well as small pale-green limpets. The ROV lifted off again and two hydrothermal vent chimneys came into view, sitting on a platform of craggy basalt, crowded at the base with the flabby anemones, more clumps of barnacles and patches of snails and crabs. The first chimney was a dark rust colour, looking much like a dirty old pipe, with a scattering of barnacles up its entire length, probably 3–4 metres high. The second chimney, immediately adjacent to the first, was slightly larger and bright white, crowded with Hoff crabs and anemones around the base and pouring clouds of what looked like billowing black smoke from multiple pinnacles at the summit. In places on the East

Scotia Ridge we found the temperature of this 'black smoke' was up to 386°C. Towards the top of the chimney the largest yeti crabs, all males and draped in stringy grey mats of bacteria, climbed about, occasionally battling one another, the victor dislodging his opponent and sending him tumbling down to the seabed metres below. I watched in fascination as one got too close to the billowing vent fluid, suddenly jerking back on its hind limbs to avoid getting burnt.

Up on the ship we examined the yeti crabs. They were amazingly tough, and despite being hauled up through over 2,000 metres of water, a pressure change of 200 atmospheres, they survived in aquaria for several days. The large males, draped in their bacteria cloaks, resembled old druids, stank of rotten eggs – the signature aroma of hydrogen sulphide – and were covered in bacterial slime. Burns were visible on their limbs showing that sometimes the crabs got too close to the superheated vent fluid. The female yeti crabs tended to live a short distance away from the vent chimneys especially if they were carrying eggs. Sven Thatje, a tall German scientist working at the University of Southampton, would appear from time to time from a refrigerated aquarium he was working in to report to us his latest observations. Every time he appeared I could not shake the impression of a young Dr Frankenstein coming to report on his latest sinister experiment ('It's alive!'), but what he was finding was extremely interesting. The crabs held their eggs until the larvae were developed to a very advanced stage before they hatched and swam away. This suggested that the larvae did not spend much time in the water before settling down on a vent to grow. We had seen numerous miniature yetis crawling around among the stalked barnacles around the vents. This raised the question of how these animals migrated from one vent to another. Of course, the answer was that the surrounding waters of the Southern Ocean are extremely cold, meaning that even if a larva was released at an advanced stage it might last in the Antarctic Circumpolar Current for a long time and travel a long distance as it would grow very slowly.

The density of Hoff crabs around the vents was staggering but was testament to the high productivity achieved by bacterial chemosynthesis in fluids rich with hydrogen sulphide. The Hoff crabs actually farm sulphur bacteria on the hairs on the underside of their bodies. The crabs comb the bacteria from these hairs using their claws and eat it. And, as we saw, some of the large slimy male yetis found on the vent chimneys were draped in bacteria. The reason for so much argy-bargy where vent fluids emanated from the seabed, especially in areas of diffuse flow, was because the yeti crabs were trying to find the best conditions to allow their bacteria to grow. However, too close to black smokers, where the flow is much more focused, the conditions become too harsh for the smaller juvenile and female crabs, so only the large male 'beach masters', as Nicolai nicknamed them, can survive.

The Hoff crabs, though, lived on a knife edge. They were not seen outside of the vent environment, and we knew that experiments had suggested that large crustaceans shut down at very low temperatures because of a failure to regulate magnesium ions in the body fluids. This explained why crabs and lobsters, one of the dominant groups of shell-breaking predators elsewhere in the ocean, were generally missing from the Antarctic. The Hoff crabs could survive only around the vents, although their larvae must have been able to survive beyond to migrate to other vent sites. Around the vents themselves the yeti crabs were preyed upon by the large sea anemones and roving hydrothermal vent octopuses, which were a pale-grey colour with sinister dark blue-grey eyes. Around the periphery of the vents was a ring of seven-armed starfish that we also saw preying on any yeti crabs that happened to stray a bit too far from the warm waters emanating from the seabed. The starfish completely enveloped the yeti crabs, forming a dome over them with their arms and no doubt digesting their victims externally. No one on board the ship could fail to be charmed by these blind furry crabs living in the pitch-dark on the island-like oases of warm water, jostling with each other for the best spot around the vents. After the cruise, John Copley had a human-sized yeti crab suit made for education events – needless to say the children adored it and the students had great fun impersonating the Hoff.

The atmosphere on the ship for the six weeks we were at sea in January to February 2010 was electric, like being at a football match for the entire time as discovery after discovery was made. From the East Scotia Ridge, we steamed over to the South Sandwich Islands where on the previous Antarctic cruise we had found a new submerged volcano sitting next to Kemp Seamount. The floor of the crater was nearly 5 kilometres across covered in fine mud grazed by sea cucumbers but also had a small sub-cone where volcanic activity was still present. This mini-volcano comprised the strangest sub-sea landscape I have ever seen. There were complexes of white chimneys, some formed of sulphur or carbonate so delicate that they resembled the spires of fairyland palaces from the wildest imaginings of the Brothers Grimm. These vents received names like 'Winter Palace', 'The Great Wall', and, my favourite, Disney's 'Toxic Castle'. The vents were releasing white clouds of fluid, an appearance caused by lower temperatures than that of the black smokers, generally of 200°C or less, which gives rise to a very different chemistry.

This dreamlike landscape was more like a forbidden garden, though. Lying around the vents and crevasses were piles of dead squid, shrimps and other deep-sea life being feasted upon by large numbers of small sea spiders. Although related to land spiders, they belong to a different class called the Pycnogonida. They look like spiders made out of pipe cleaners with slender cylindrical bodies so thin that their body organs poke down into the upper parts of their sticklike legs. These animals are usually specialist predators of sea anemones, sponges and other marine invertebrates that live fixed to the seabed. At the hydrothermal vent crater they had assumed the role of scavengers. So toxic were the waters around the vents that we even witnessed, via the cameras on *Isis*, a squid swim into the vent field and immediately become paralysed and crash to the seabed like a plane that had been shot down. Later, the chemists on the cruise, led by Doug, discovered very high concentrations of hydrogen sulphide, hydrogen fluoride and other toxic chemicals in the fluids, which were highly acidic. Despite this, not only sea spiders managed to live in this environment, but also limpets crawling over rock made of sulphur and the anemones familiar from the East Scotia Ridge. Elsewhere we found large clams among broken boulder fields of basalt and also living in ash and sand, visible by their pipe-like siphons used to draw water in and out of the buried animals, allowing them to 'breathe'.

One of the most memorable incidents for me, however, was when *Isis* was travelling up a chute or a slope on the flank of the volcano. It was covered in a pale-green, smooth, almost gelatinous material. I stared and stared at it, as did the others in the ROV control container. Was it a mineral or was it a mat formed by bacteria or another type of microbe? The substance was so odd-looking that there was no frame of reference to compare it to. Guesses were thrown around in the confined space of the control cab, but none of us could agree on what it might be. To look at something on Earth, and as a scientist simply have a question mark appear in your brain, is a very special feeling and one rarely experienced on land because it is so well explored. I savoured the feeling because it was so unusual and mentally challenging. Here, out in the unexplored depths of the Southern Ocean, it was quite possible to find something beyond the realm of human experience.

On the slopes of the sub-cone we also discovered the skeleton of an Antarctic minke whale. This comprised the vertebrae, the skull, some ribs and the whale equivalent of the shoulder blades. Despite the constant attentions of a large red squid, attracted by the lights of *Isis*, and the odd fly-past by a large grey Antarctic toothfish, we filmed, mapped and sampled the remains. This was only the sixth natural whale fall found in the world. In the high-resolution cameras of *Isis*, we could see small crustaceans, worms and limpets crawling over the corroded yellow-white vertebrae. From the bones also blossomed dense strands of trans-lucent tubes with a red blood vessel through the centre and ending in a plume of fine tentacles. These, we knew, were zombie snot worms, a recently discovered animal that survives by absorbing the oils from the inside of the whale bones through a ramifying network of tiny tubes. Zombie snot worms are only found living on whale bones, and they have symbiotic bacteria that convert the oils to food. The worms are all females as the males are tiny and live parasitically on the females. I spent a long afternoon in the cold-room of the ship trying to extract snot worms from a vertebra we had picked up. This turned into a rather frustrating and difficult task, as once the worms were prised from their bony burrows they did, indeed, resemble a blob of snot. In the end I resorted to trying to cut the bone in half with a hacksaw. The cold-room where this operation took place was rapidly evacuated as the smell was deeply unpleasant, a cross between a very ancient tannery I had once visited in Morocco and burning hair. The smell was incredibly pungent and took days to scrub off my hands and out of my hair. We did extract some of the worms eventually and they comprised three species, two of which were new, one of which was named after me as *Osedax rogersi* by colleagues at the University of Southampton and the Natural History Museum. So yes, I ended up having my name attached to a zombie snot worm – something for posterity, no doubt.

During the 2010 expedition to the East Scotia Ridge and South Sandwich Islands, almost every species found on the hydrothermal vents was new. This included the yeti crabs, now named as *Kiwa tyleri*, the snails, *Gigantopelta chessoia*, and the barnacles, *Vulcano-lepas scotiaensis*. Every one of these species told its own story about the evolution of life in the ocean. For example, as new species of yeti crabs have been discovered in the deep sea, four more since our discovery of *Kiwa tyleri* less than 10 years ago, a combination of studies of fossils and DNA sequencing has allowed us to pinpoint when the crabs first appeared. Nicolai Roterman, at the time of writing this a postdoctoral scientist at the University of Oxford, has discovered that, contrary to first contentions that the vent fauna was very ancient and somehow stable despite changes in the Earth's environment, the yeti crabs evolved as recently as 30 million years ago somewhere in the south-east Pacific. This coincided with a time of cooling in the deep ocean when oxygen concentrations increased. Prior to this, a sudden climatic warming event known as the Palaeocene-Eocene Thermal Maximum (PETM) may have had severe consequences for animals living around vents as because vent fluids lack oxygen in their undiluted state they are highly dependent on oxygen from the surrounding ocean. The role of oxygen in evolution of ocean life will be discussed later, but suffice it to say that vent animals may be highly vulnerable to climatic disturbances, despite living in the deep sea, because of their reliance on oxygen. Studies on other groups of vent animals, including the stalked barnacles we found in the Southern Ocean, support such a recent origin for this special fauna.

The extreme novelty of the East Scotia Ridge hydrothermal vent fields was also a surprise. If our theory that the Southern Ocean was a 'highway' for species to travel between the Pacific, Indian and Atlantic Oceans was correct, we should have found species from these regions on the East Scotia Ridge. These included animals like tube worms, mussels and vent shrimp. Although none of these groups were found, the limpets on the East Scotia Ridge vents were very closely related to species from the Atlantic and the clams from the South Sandwich Islands crater resembled a widespread group associated with chemosynthetic ecosystems. Everything else, at the time, was new. Nowhere had the spectacle of such dense, heaving aggregations of yeti crabs been seen.

At best it seemed that the Southern Ocean acted as a selective gateway for the migration of species over large distances, but one which most of the vent species familiar from other oceans could not cross. The clue as to why this was the case might lie in the life histories of the animals in question. It is known that the extreme seasonality of the Antarctic, where much of the Southern Ocean is dark and covered in sea ice over winter, coupled with low temperatures, selects against animals that have young or larval stages which feed on the plankton as is difficult for them to survive in such conditions. Many of the missing groups displayed this form of life history. The Southern Ocean vents really were unique, being typified by the large numbers of yeti crabs and vent snails, which were not seen elsewhere. Our single expedition had given clarity to how animals were distributed on vents globally and we estimated that there were 11 distinct communities of animals found on hydrothermal vents around the world. This finding still stands today.

The question 'Is there life out there?' has entranced scientists and science fiction writers for centuries, but the finding of hydrothermal vents spawned the science of astrobiology. The discovery of hyperthermophiles, organisms such as vent bacteria that thrive at high temperatures, widened our evaluation of the conditions where life might occur elsewhere in the universe. The so-called Goldilocks zone, the distance from stars where the conditions suitable for life may be found, expanded. The discovery of chemosynthesis, and that life may have originated in hydrothermal systems, broadened our ideas about where life may exist even further and most recently has drawn our eyes back to our own solar system. Could there have been a second genesis of life around the sun?

At present, the search for life elsewhere in the universe is hampered by the astronomic distances between our solar system and other stars and their orbiting planets. Evidence for planets around distant stars, including planets in the Goldilocks zone, has rapidly accumulated over the last decade. However, we simply have little idea of the chances of genesis. Was Earth a lucky fluke, repeated but very rarely throughout the universe? Alternatively, is it the case that where a planet lies in the right location around a star, and hydrothermal vents or hot springs exist along with the presence of liquid water, life is 'inevitable'? If the latter is right the implication is that life among the stars is a frequent occurrence. The exploration of our own solar system has thrown up several possible candidates that may host hydrothermal vents. The most likely are the icy moons of Jupiter and Saturn, including Ganymede, Callisto, Titan, Europa and Enceladus. The latter two moons are probably the best studied in the context of the possibility of a second genesis of life in the solar system.

# Deep-Sea Hot Springs

The accidental discovery of hydrothermal vents and the communities of life around them by deep-sea biologists and geologists has completely changed the way we think about life on Earth and elsewhere in the universe. As a scientist these environments are fascinating but I am left wondering what other remarkable discoveries are waiting out there in the deep ocean. However, we might not have long to explore this remote and mysterious wilderness before industry leaves its mark. As I am writing this chapter, Japan has announced that it has undertaken the first mining of hydrothermal minerals from the deep sea off Okinawa at depths of 1,600 metres. Hydrothermal chimneys, like those we discovered in the Southern Ocean, and their surrounding vent deposits are rich in metals – in particular, copper, but also zinc, gold, silver and other rare minerals.

When we recovered samples of chimneys from the East Scotia Ridge the insides were encrusted with scales of bright golden copper pyrite, an ore of copper formed by copper, iron and sulphide. In a world where new technologies are placing ever increasing demands on such minerals, marine mining is a real prospect. Nautilus Minerals, a Canadian mining company, is planning to begin mining hydrothermal vents off of Papua New Guinea, at similar depths to Japan, in 2019. Twenty-nine licences have been granted by the United Nations International Seabed Authority (ISA), to explore areas of the high seas for minerals, including one right next to the Lost City vent site. The race to the deep has begun before we even know what other life and what other ecosystems are present.

My experience of other industries in the deep ocean is that

exploitation ramps up quickly, before scientists have had the opportunity to investigate the potential consequences for the environment. Money and profit still come before science and stewardship of the ocean, and this is particularly concerning when we know so little about the deep sea. In many ways these industries take advantage of the fact that what they do is often far from land and therefore difficult to monitor. In many cases governments are complicit in these activities. Many times at intergovernmental meetings have I heard delegations ask the question: 'Where is the evidence that this is damaging the environment?' or 'What is the scientific evidence that this ecosystem should be protected?'

They know full well that neither they, nor anyone else, has stumped up the funding to look at the impacts of their activities on the species, habitats or broader ecosystems on which they are targeted. The two industries I have had direct contact with in this context are the offshore oil and gas industries and deep-sea fishing. This is where I will take you next, surprisingly close to home off the coast of Britain and Europe.

# 3

# Protecting the Gardens of the Deep:

*Greenpeace vs The Secretary of State*
*for Trade and Industry*

Every day I cycle to work at the University of Oxford not quite knowing what will turn up. There might be an email asking me to assess the latest efforts to sustainably manage deep-sea fishing, or a call to discuss the latest moves at the United Nations (UN) over negotiations to conserve diversity in the high seas. Yesterday there was an enquiry from BBC Radio 4 asking about the feeding habits of six-gilled sharks in the deep sea. Then there are the usual duties of a university professor: lecturing, giving tutorials, setting exams, supervising student projects, attending college meetings and running a large research team. All this adds up to several jobs' worth of work, the hours are long and, as the words of a song by Elbow relate, I struggle with deadlines like cats in a sack. My family: my wife, Candida, and my two daughters, Zoe and Freya, suffer for my work. When I'm home I'm often in the office until late, and then there is the fieldwork, when I will be away for up to six weeks with little communication. The ocean, that wonderful place, home to coral reefs, manta rays, whales, turtles, sharks and a dizzying array of other species, is in serious trouble. Each morning I rise with an undiminished determination to fight the ocean's corner using the tools that I have to hand, my science. All of my work, my research, the policy advice I give and the public outreach I do is aimed at waking the world up to the plight of the ocean today and to enable us to do something about it.

Of course, this process is collaborative and involves a truly global effort. In 1999, I was a witness for the environmental charity Greenpeace in a high-profile court case. I saw first-hand the extent to which industries in the deep sea turn a blind eye to the need to scientifically explore the ocean further and the true impact their operations are having. I had never acted as an expert witness before, so providing a legally sworn document to Greenpeace as evidence in a legal action against 10 oil companies, as well as the UK government, was a baptism of fire. The judicial review was focused on the European legislation that dealt with habitat conservation and its application in the deep sea. Arriving at the Royal Courts of Justice, a monument to Victorian Gothic architecture in London, was an intimidating experience. It's a large building constructed of pale-grey stone with great arched gates at the front flanked by towers and arcades of lancet windows. To my eyes it was like a paler, less elaborate cousin of the Houses of Parliament. I was ushered in by Debbie Tripley, the lawyer for Greenpeace, and met with the Queen's Counsel, Nigel Pleming. He was a tall man, dressed in the black gown of his profession.

'Welcome, welcome!' he said, smiling confidently and shaking my hand after Debbie had introduced us. 'You've done a great job. Everything is going to go well today, I'm sure of it.'

I'd wished I'd had his confidence in that moment, but truthfully I was beyond nervous. Greenpeace had invited me to see the case being heard and I had been keen to see how my evidence was received but could only attend the first day because of work pressures.

Debbie spoke in hushed tones: 'When you go in, you'll be sitting in the rows at the back. The Queen's Counsel and lawyers will be in front of you. There is no need to be nervous. You have sworn an affidavit that your evidence is true and that has been accepted so you will not have to stand up and say anything.'

'Right,' I said. It was just about all I could manage.

'The judge, the Honourable Mr Justice Maurice Kay, is thorough.'

'Is that a good thing or a bad thing?' Having never been in court before, and assuming that the dice would be loaded against Greenpeace, I genuinely had no idea.

'It's good. Your evidence is detailed and it's strong. It's based on the latest science and he *will* read it, so we'll be fine.'

Before I realized it, we were being led into the courtroom and I shuffled down the bench and took my place. I was trembling slightly with adrenaline. The room was poorly lit and panelled in dark wood. I was looking over the shoulders of rows of lawyers, dressed in black gowns with grey wigs. Debbie sat next to me and leaned over.

'There's Nigel and our team,' she whispered, indicating the Greenpeace QCs. 'In front of us are the lawyers on the other side. Oh look, they seem to be taking a great deal of interest in your report,' she said delightedly.

'That's the last thing I need to see,' I said nervously, only half-joking.

I immediately caught sight of the two lawyers. It was unclear whether they were representing the Secretary of State for the government or the oil companies, but they were poring over the report I had written. My report focused on whether corals could form reefs in the deep, cold waters of the north-east Atlantic and where they occurred with respect to proposed oil exploration areas off the UK coastline. The judicial review focused on whether the European Habitats Directive should apply to the offshore seas of the United Kingdom. The Habitats Directive was one of the cornerstones of European legislation aimed at conserving threatened or rare species and habitats against potentially damaging development, whether it be industrial or otherwise. At the time, the regulations only applied to 12 miles offshore but it had not been established whether they also applied to the entire exclusive economic zone of the UK, which extended to 200 nautical miles from the coast, including large areas of deep sea. The Directive required that habitats of conservation importance should be protected. Reefs were classified as having conservation priority under the Habitats Directive because of their high richness in associated species. What was at stake was whether the UK government would be placed under a legal duty to exclude oil production and other industrial activities from areas of the seabed where such habitats named in the Directive were found.

As I watched, I saw the two lawyers pointing animatedly to something in my report. I went cold and my hairs stood on end. What were they pointing at? I could see the section they were reading and I was desperately thinking about what was written on those particular pages. The lawyers prodded someone in front of them and there was an excited exchange.

Doubts nagged at me. I'm not the type of scientist with an unbearable self-confidence in everything I do. My background as a first-generation academic from a working-class family forbade that kind of arrogance. I started to question myself. Had I got everything right? Were there mistakes on that page that they could seize upon and that would enable them to tear apart my evidence? My reputation as a scientist was very much on the line. To make a scientific error in such a high-profile court case would ruin my credibility as an expert and likely result in accusations of bias by those who would profit from Greenpeace losing. I have seen industry representatives try to discredit scientists many times to undermine their scientific work and evidence. It would prevent me from taking on other industries harming the ocean in the future and affect my career as a university academic. Suddenly, the proceedings began and a hush fell over the court room like a blanket as the judge spoke.

The judge introduced everyone to the proceedings and Nigel stood up to give an account of Greenpeace's evidence. Whenever a point was scored with new evidence about deep-sea corals or the scope of the Habitats Directive, the other wigged lawyers on the Greenpeace team grinned happily, leering over at their opposition. This was a piece of theatre I wasn't expecting, more Punch and Judy than the High Court of the land. In any case, I was far too tense to enjoy this psychological warfare as the discourse twisted one way and then the next. My evidence that corals could be severely damaged by oil spills – something that was to be proven many years later following the *Deepwater Horizon* tragedy, where a deep-water drilling rig in the Gulf of Mexico suffered a blow out and exploded, leading to the largest marine oil spill in the history of the oil industry – was described to the judge by the QC, Nigel. As he spoke, a scene not unlike one from a TV drama seemed to unfold: a government scientist rushed down to the front of the court proclaiming he had evidence that the main deep-water reef-building coral, *Lophelia pertusa*, could grow on the legs of oil platforms. I stopped breathing. Was this the point at which the case would unravel? It was well known to deep-sea scientists, including myself, that this coral could grow on submerged artificial structures, but that did not mean it was somehow immune to the toxic effects of oil or the chemicals that might be used to disperse it in the case of a spill. Luckily, the judge off-handedly dismissed this interruption as being irrelevant to the case. The judge had been correct in that, regardless of whether the coral was capable of forming colonies on artificial structures, it was the ability to form a reef habitat that was the subject of contention.

Hours later I left the court exhausted even though I had only sat and watched the proceedings, but Debbie came out smiling.

# Protecting the Gardens of the Deep

'I think it went well,' she said. 'At least the case is going to be heard in its entirety and not dismissed on a matter of timing like the last time.'

I headed back to Southampton, where I was working at the time, as the case continued to be heard throughout the week. As a scientist I was very focused on the technical aspects of my contribution to the judicial review but I was not aware of the full implications of the case or the turbulent history between Greenpeace, the UK government and the oil companies operating in the North Sea and the Atlantic, west of Scotland.

This began in April 1995 when Greenpeace activists had boarded an abandoned oil storage platform, the *Brent Spar*, to protest at its disposal in the deep sea to the west of Scotland. The *Spar*, which was jointly owned by Shell and Esso, was one of the first installations in the North Sea and would need to be removed as oil and gas production wound down in the coming decades. The option of deep-sea disposal of a 14,500 tonne structure had largely been decided upon by Shell, and put forward quietly to the UK government as the safest and cheapest route for disposal. Shell would receive a major tax break for the costs of disposal, and economic considerations were significant to the Conservative government of the time, which was led by John Major. A Greenpeace campaigner, Simon Reddy, with whom I was to work many years later at the Global Ocean Commission in Oxford, discovered the plan for disposal from an industry contact. Greenpeace's occupation of the *Spar* was followed closely by the media, while claim and counterclaim were made by the campaigners, Shell and the UK government. The UK Minister for Energy at the Department of Trade and Industry, Tim Eggar, claimed that the dump site, which was at 2,000 metres depth in a location called the North Feni Ridge, contained 'just a few worms'. This contention was based on the environmental reports used by Shell, which in turn were based on an analysis of the deep ocean at abyssal depths (4,000 metres or more) by an environmental consultancy commissioned by the oil company.

Several prominent UK deep-sea scientists severely criticized the Shell work, including John Lambshead from the Natural History Museum, and John Gage and John Gordon, both from the Scottish Association for Marine Sciences (SAMS) in Oban, Scotland. These scientists, all world leaders in their fields, had not been consulted by Shell or the UK government on the disposal sites proposed for the *Spar*. Shell's information was out of date.

In the late 1960s, deep-sea biologists had discovered that the depths between 1,000 metres and 3,000 metres, in what was termed the bathyal zone, was extremely rich in species. These were mainly small animals living on or in the seabed, including worms, crustaceans, snails and clams. Work published by John Lambshead, the same year as the *Spar* affair, showed that the diversity of species living in the seabed off Rockall grew with increasing depth to about 1,500 metres and then slowly decreased as the water got deeper still. Claims that the site in which the *Spar* was to be dumped was a biological desert were clearly wrong, but it was an argument that I was to encounter repeatedly in clashes with the deep-sea trawling industry in the years to come. Greenpeace also made mistakes, claiming that the *Spar* contained nearly 5,000 tonnes of waste oil as well as a secret stash of toxic chemicals. These claims were based on technical errors made in measurements of the oil remaining in the storage tanks of the *Spar* and on poor advice that Greenpeace had received. In fact, it was estimated that there was likely 74–103 tonnes of waste oil in the rig, up to a maximum of 130 tonnes, and the hidden toxic waste did not exist.

Following an increasingly hostile reaction to Shell's dumping plans and growing pressure across Europe to ban deep-sea dumping of oil installations, the decision to dispose of the *Spar* at sea was reversed in June 1995. Shell's change of heart was a total shock to the UK government of the time. There followed a back-lash on Greenpeace from the UK government that they had made wild allegations that were not based on fact and these had carried the day. The media also back-pedalled, accusing Greenpeace of misleading them. The mistakes Greenpeace had made during the campaign, which they had freely admitted to as soon as they had come to light, were seized upon. The UK government and the oil industry fought against the prevailing view in Europe that dumping at sea was no longer acceptable. On this issue, the UK was very much the dirty man of Europe. However, later that year European ministers at the Oslo-Paris (OSPAR) convention in Esbjerg, Denmark, declared a moratorium on deep-sea dumping, and in July 1998 at a meeting in Lisbon, this moratorium was turned into a ban of at-sea dumping for most offshore structures.

As Greenpeace claimed victory over the dumping of off-shore structures in European waters, a new debate had emerged concerning the oil and gas industry. In 1990 and 1995 the Intergovernmental Panel on Climate Change (IPCC), a body of scientific experts representing many countries in the UN, had sounded the alarm over $CO_2$ emissions resulting from, among other things, the burning of fossil fuels. This laid the ground for the Kyoto Protocol, adopted in 1997, an international agreement for countries to reduce their $CO_2$ emissions. Greenpeace responded to the finding that the burning of fossil fuels was leading to dangerous climate change by embarking on the Atlantic Frontier Campaign, the aim of which was to persuade the UK government to invest in renewable energy rather than open up new oil production opportunities to the west of Scotland.

The very public critique of Shell's environmental assessments by leading deep-sea scientists had revealed the richness of deep-sea ecosystems around the Rockall Bank. When I was first contacted by Greenpeace in 1997, they asked me to analyse the distribution of coral in the area designated for oil exploration to the west of Scotland. They also expressed an interest in whether or not the coral could form 'reefs', which was important because this was one of the named habitats that required protection under the European Habitats Directive, which defined them as follows:

*Submarine, or exposed at low tide, rocky substrates and biogenic concretions, which arise from the seafloor in the sublittoral zone but may extend into the littoral zone where there is an uninterrupted zonation of plant and animal communities. These reefs generally support a zonation of benthic communities of algae and animal species including concretions, encrustations and corallogenic concretions.*

# Protecting the Gardens of the Deep

Answering the question of whether *Lophelia pertusa* could form reefs in the deep sea became key to the judicial review Greenpeace was to bring to the High Court. If it was proved that habitats of conservation priority under the Habitats Directive occurred in the deeper waters under the jurisdiction of the UK government then it followed that there was a legal duty to protect them.

*Lophelia pertusa* was first described in print by the Bishop of Bergen in 1755. The bishop had a sideline in natural history and in his book *The Natural History of Norway* he described a particularly fine white specimen of the coral with flowers. Carl Linnaeus entered the species as *Madrepora pertusa* in his *Systema Naturae*, the work that was to set the path for modern classification of life on Earth. In the late nineteenth century the British naturalist Philip Henry Gosse illustrated the coral in his book on sea anemones, *Actinologia Britannica*. There followed descriptions of coral habitat lying in the deeper waters off Europe by other natural historians. As late as 1948 the French marine biologist Édouard Le Danois described coral 'massifs' on the continental slope west of Ireland and along the margin of the Bay of Biscay. His book *Les Profondeurs de la Mer* (*The Depths of the Sea*) showed maps of where the coral banks were along the coasts of Spain, France and Ireland. Yet, somehow, the presence of these corals had largely been forgotten just decades later. Why this was the case is unclear but may have simply been because reefs generally occurred in rocky, steep areas that were hard to sample using conventional, over-the-side sampling gears such as trawls, dredges and corers, the latter used to take quantitative samples of deep-sea mud. From the 1960s to the 1980s these were the main tools of deep-sea science. Deep-sea research was focused on understanding the patterns of abundance and diversity of species of animals living on or in muddy or sandy habitats which comprised much of the deep seabed. *Lophelia pertusa* or other deep-water habitat-forming corals were certainly not a topic of focus when I studied for my degree and PhD at the University of Liverpool's Port Erin Marine Laboratory on the Isle of Man in the late 1980s.

Ironically, it was the oil industry that 'rediscovered' deep-water coral after sponsoring research to understand the geology of the edge of the European continent, a zone called the continental slope. To the west of Britain this is particularly complicated by a series of banks and seamounts. These are the remains of the tectonic convulsions that accompanied the start of a new mid-ocean ridge or submarine mountain range to the west of Europe during the time of the dinosaurs. The ridge subsequently died (along with the dinosaurs) but left behind a legacy in the form of a cluster of submarine volcanoes including the Anton Dohrn Seamount and Rosemary Bank, as well as the Rockall Bank itself, which is a fragment of the European continent. Lumps and bumps on the seafloor, from the smallest submarine hill to seamounts thousands of metres in elevation, provide good habitat for corals and other fauna that live fixed to the seabed. This is because they obstruct ocean currents flowing past them, causing the currents to accelerate and also generating eddies and vertical flows of water moving upward or downward. Strong currents bring lots of food for corals in the form of zooplankton and particles of organic material, as well as sweeping away sediments to reveal hard rock, which the animals need to attach to.

# Protecting the Gardens of the Deep

*Lophelia pertusa* is the most conspicuous coral living in deep waters off the European continent. This species, like shallow-water reef-forming corals, is a stony coral. It forms branching colonies from the size of a cauliflower to a bushy thicket, which range in colour from a beautiful pure white to pinkish or orange. The 'polyps' – the living parts of the coral – resemble small anemones or flowers up to a centimetre or so across, which comprise a mouth surrounded by translucent tentacles, pixilated with tiny pale spots. The polyps live in cup-shaped structures called 'calices' reinforced by thin, radiating rays of skeleton, which, along with the branches, are made of a form of calcium carbonate called aragonite. It is extremely hard and feels like stone but is also very brittle, especially when the branches of coral are thin. Unlike the reef-forming corals of shallow waters, *Lophelia* does not carry symbiotic algae (or 'zooxanthellae') in its tissues – this algae being what supplies tropical corals with most of their food through the process of photosynthesis. This is because *Lophelia* lives in the dark, cold waters of the deep sea at depths of 1,000 metres or more. The absence of bright sunlight means that photosynthesis used by the symbiotic algae to generate carbon-based or organic molecules is not possible. Instead, colonies of *Lophelia* can be thought of as walls of mouths, with each individual polyp extending its tentacles to capture live zooplankton, the tiny animals and larvae that drift in the ocean, which they sting to death with their 'cnidae', stinging cells possessed by all corals as well as their relatives, the sea anemones and jellyfish. As a child in the rock pools of Sligo, I used to brush my fingers against the tentacles of a bright-green anemone called a snakelocks. They would stick to me because of the stinging cells firing into the skin but not deep enough to deliver any venom. In *Lophelia* the stinging cells explosively release a tube covered in spiralling rows of spines which penetrate the tissue of prey and deliver a deadly venom. These microscopic barbed lances are also too small to penetrate human skin in *Lophelia* but those possessed by tropical jellyfish can certainly cause serious injury or even death to humans through delivery of potent toxins that can cause agonizing pain. So although corals like *Lophelia pertusa* form beautiful structures which are pleasing to the human eye it is important to bear in mind they are efficient predators that lie in wait for their prey in the deep, dark waters surrounding our continents, submarine canyons and mountains.

To me, the most logical way to proceed in my quest to discover whether *Lophelia pertusa* could be regarded as reef forming was to compare it with tropical coral reefs. The lack of symbiotic algae was a major difference between *Lophelia* and its shallow-water cousins. However, like shallow-water reefs, *Lophelia* colonies could coalesce over time to form massive structures elevated from the seabed, in other words, reefs. The biggest of these, discovered as I was doing my research for Greenpeace, was the Sula Ridge coral reef off Norway. Here *Lophelia pertusa* formed a structure up to 35 metres in height, 13 kilometres long and 700 metres wide. Later the Røst Reef was discovered off the Lofoten Islands in Norway, 45 kilometres long and with a width of up to 3 kilometres. To the west of Britain large coral mounds were discovered off the Rockall Bank and also in a large indentation into the continental slope off the south-western tip of Ireland called the Porcupine Seabight.

# Protecting the Gardens of the Deep

In 1973, a geologist called John Wilson undertook the first dives on the reefs in a submersible called *Pisces III*. Black-and-white film from the submersible, which is still available online, showed swathes of bright-white coral bushes with polyps extended fading into the darkness. I met John several times. He was a small and very lively man with white hair and a beard and he spoke in a soft Scottish accent. He came up with a theory of how the deep-water reefs formed. A colony grows on the seabed and as time progresses it gets larger but is also colonized by animals that eat away at its skeleton. These colonists are called bioeroders and include sponges, various segmented worms, clams and other animals. The sponges etch away at the coral using chemical attack, producing tiny chips of coral skeleton that form carbonate sands or finer silts. The worms attack the coral with their jaws. The result of this constant nibbling away of the coral skeleton is that pieces of the coral, with still-living polyps, fall off the parent colony on to the seabed forming a ring of satellite daughter colonies, which then grow. These daughter colonies endure the same nibbling as the parent and eventually a ring-like structure is formed, the initial colonies becoming choked by the growing framework and dying. These structures, named 'Wilson Rings' after John, coalesce to form increasingly large frameworks. Coral grows on dead coral and because water-flow through the framework slows down, sediment becomes trapped within it. Eventually a complex three-dimensional structure is formed, elevated off the seabed comprising mainly of sediment and dead coral skeleton but with a rind of living coral.

So, like shallow-water reefs, deep or cold-water reefs formed by *Lophelia* can form massive and highly complex structures. I began to count the numbers of species that lived with *Lophelia pertusa* in the gloomy depths of the north-east Atlantic. To my amazement I found nearly 900 species lived on cold-water reefs off the coast of Europe and in the Mediterranean. These included a very wide range of animals. Sponges were particularly diverse, from tiny species boring into the coral framework to massive white, orange or yellow colonies growing on top of the corals. Some, like *Aphrocallistes*, had delicate glass skeletons resembling vases formed by the most delicate vitreous filigree. Hydroids were also diverse, a group of encrusting animals related to corals and jellyfish that varied in appearance from a fuzz growing on coral skeleton to erect ghostly white miniature Christmas trees. Other corals, such as sea fans and bubblegum corals – so named because they look like huge rose-pink trees with blunt-ended knobbly branches resembling lumps of bubblegum – were also common inhabitants of *Lophelia* reefs. This was diversity on diversity, a fractal world where the closer you looked the more life was discovered.

Not all groups on *Lophelia pertusa* reefs were as diverse as on tropical coral reefs. Fish, for example, had a much lower diversity than tropical shallow-water reefs. Anyone who has dived on a tropical reef cannot help but be bowled over by the abundance and variety of fish living in the corals, swimming among coral branches or travelling around the reef. My first experience of this had been in the Maldives on honeymoon. The clear blue waters revealed a stunning multicoloured garden of staggering complexity with corals, sponges, sea fans and soft corals coming in a bewildering array of shapes and textures. All swarmed with fish, from small damselfish patterned like mint humbugs to shoals of yellow grunts and larger black-and-white striped sweetlips with yellow black-spotted tails. Sharks, rays and groupers patrolled the reefs and if you were lucky you might see a shoal of large grey bump-head parrotfish. Tropical coral reefs are home to a quarter of all marine fish species. The cold-water reefs formed by *Lophelia pertusa* were inhabited by relatively few fish species, reflecting the lower diversity of the north-east Atlantic at the depths where they were found.

Another large difference between *Lophelia* reefs and tropical reefs was the diversity of the reef-forming corals themselves. I discovered that there were deep-sea coral reefs formed by other coral species around the world, but the primary reef-former was always one species. A tropical shallow-water reef is formed of myriad coral species. There will be tables of spiny corals, great trees like deer's antlers, brain-like massive corals, lumpy mounds, stove-pipes, mushrooms; the variety bewilders the mind. This is because the reef-forming corals we have today in shallow waters have been evolving for more than 200 million years, since the time the dinosaurs first appeared on Earth.

One of the other features of shallow-water coral reefs I needed to consider in the question of whether or not *Lophelia pertusa* was reef-forming was that of symbiosis. What I mean by this is a relationship between species where one benefits without harming the other (commensal) or where both benefit (mutualistic). Symbiotic relationships between different species are a frequent and notable feature of shallow-water coral reefs. Not just the relationship between the corals and microscopic algae, but also between other life on the reef, such as anemones and anemone fish, also known as clownfish. The anemone fish defend the sea anemones from their predators and parasites and in return gain protection by hiding from larger fish among the stinging tentacles of the anemone, to which they are immune. The fish also get scraps of the anemone's food, and waste from the fish provides nitrogen for symbiotic algae that live in the tissues of the host, as in shallow-water corals. The problem here was that *Lophelia pertusa* reefs simply had not been well-enough studied for us to understand much about whether there were such relationships between species. There is a species of shelled amoeba called *Hyrrokkin sarcophaga* that forms cyst-like structures on the surface of the corals, but this appears to be a parasitic relationship, one that harms the host. However, there was one relationship that appeared to fit the bill as a mutualistic relationship. A large, predatory worm, *Eunice norvegicus*, forms parchment-like tubes in the framework of *Lophelia* reefs. The corals secrete aragonite over the tubes making the worm tubes hard and robust and overall the tubes strengthen the whole coral framework. Larval coral settle on the worm tubes and grow, so the worm actually helps the reef to form – a relationship benefiting both animals. What's more, *Eunice* will viciously defend the coral from any predator potentially interfering with its patch of reef.

I witnessed this defence in the flesh almost 20 years ago, when I was on an expedition to the northern Rockall Trough on the ship RRS *Discovery*. Paul Tyler, my friend and colleague from the University of Southampton, and I had managed to sample some live *Lophelia pertusa* for genetic studies, and we were looking at our treasured prizes, which were being held in a large fibreglass tank on the ship. We spotted movement in one of the fist-sized fragments of coral and Paul picked it up. Out came a large, slightly iridescent, pearl-white segmented worm about the thickness of a pencil. Black eyes were visible on a head that was framed in slender tentacles. As we watched, fascinated by the complex behaviour we were seeing from such a 'simple' invertebrate defending its host, the entire front of the face of the worm erupted into a large proboscis covered in small black spines with a pair of pincer-like jaws which shot out and locked on to the soft tissue between the fingers of Paul's hand, causing him to yelp, 'Ow! The so-and-so bit me'. It was like watching the famous predator from the film *Alien* in miniature. It was a spectacular specimen, clearly capable of defending its patch of coral even from something the size of a human. The worms themselves gain protection from the aragonitic coating the corals provide for its burrow. They also steal some of the coral's food in return for their activities as a mini, spineless Rottweiler so they can also be classified as a kleptoparasite.

For me the multiple lines of evidence about the ecology of *Lophelia pertusa* gathered during long hours of research, often late into the night, led to an inescapable conclusion. The massive structures the coral formed on the seabed could only be construed as reefs. Not only did the coral build these large structures, but it also shared many other features with tropical reefs, including a high species diversity and similar processes of reef accretion and bioerosion. It was a remarkable fact that in the dark, cold waters off the European coastline these amazingly complex structures teeming with associated life had existed for thousands of years. Norway, Britain, France, Ireland and Spain had coral reefs.

Two years before the case in question, however, Greenpeace had failed in a legal challenge to the government's seventeenth licensing round where I had also acted as an expert witness. They were refused leave to move for judicial review on the grounds that they had not brought the case to the courts in sufficient time, and such a delay would lead to damage to the interests of oil companies applying for licences. The expert witness for the oil companies contested my evidence on reef forming by *Lophelia*, pursuing the line that a reef could only be such if it was a hazard to navigation. Because the case was dismissed, the question of *Lophelia pertusa* reefs was not fully examined by the court. Though we were depressed by the outcome of the case, Greenpeace did not lose heart and I decided that the best way to defend my work was to get it peer-reviewed by other scientists – the tried and tested method of ensuring that papers were critiqued by other experts in a particular field before publication.

As work progressed on the paper, scientists were finding out more and more about the whereabouts and ecology of *Lophelia pertusa* off the European continental margin. In May 1998 a cruise on the RRS *Charles Darwin*, led by Natural Environment Research Council (NERC) scientists at the National Oceanography Centre in Southampton, found a large number of mounds topped by thickets of *Lophelia pertusa* in the northern Rockall Trough. This area was included in the licensing blocks for oil exploration. The mounds lay to the south of a major barrier to the flow of water along the European continental slope, the Wyville-Thomson Ridge. They were named the Darwin Mounds, after the ship (and, of course, Charles Darwin himself).

The Darwin Mounds were of particular importance as despite their small size, a hundred metres across and a few metres high, they nonetheless fitted the European Habitats Directive definition of what constituted a reef.

With this evidence quickly coming together, final preparations were underway for the court case in London. Not only were Greenpeace presenting the report I had prepared previously, which had now been published in a peer-reviewed paper, but I had updated the information with all the new findings of the whereabouts of *Lophelia* off the coast of Britain and new discoveries regarding its biology. I also explicitly answered all of the criticisms levelled by the expert witness for the oil companies not covered by my previous reports. So it was that I found myself in the High Court of Justice on a cold morning in early November 1999.

Though I had busied myself in Southampton after my testimony at the court case, I couldn't help but think about what was happening in London. If the case went badly there would be crushing disappointment. The evidence I, and so many other scientists, had spent so long uncovering meant that, more than ever before, I felt passionately that the coral should be protected. Later in the week, the phone rang at home. I answered and was taken aback by the shouting and laughter, and what sounded like a full-scale party going on in the background. I thought it was some kind of prank call. Eventually, an elated voice floated above the mayhem and said, 'We won!'

Still confused, I muttered a baffled, 'What?'

'We won the case! We won on almost every point!' I finally recognized the voice of Richard Page from Greenpeace. Richard was an earnest and passionate campaigner for the ocean in the true Greenpeace tradition, favouring direct action. Butterflies tumbled through my stomach and I had to sit down on the stairs as the energy drained from my limbs and my hands began to shake.

'Your evidence was crucial, the judge agreed on every point. Well done! It's a shame you're not here. We'll have to celebrate when you're next in London!'

My heart was thumping, but it felt lighter and lighter as the reality of what we had achieved and the victory we had grasped sank in. 'Send my congratulations to Debbie and the others,' I managed to stutter, despite the mayhem in the background.

'I will. I will do that, and I hope you have a good evening! You deserve it!' He rang off, and so I just sat and collected my thoughts before getting up and going to tell Candida.

It transpired that Greenpeace had won the case in spectacular fashion. Apparently in response to my new evidence the oil industry expert witness had sent a simple three-line letter from a Paris hotel stating that the information he had provided in a report in 1997 remained the same today. The judge, who was clearly not impressed, explained that in view of this deficiency he had proceeded on the basis that *Lophelia pertusa* fitted within the European Habitats Directive definition of reef forming. The oil industry had taken the view that even if *Lophelia pertusa* was present in the licensing area for exploration, it was unlikely to suffer damage from oil exploration activities. This was something I had specifically addressed in my report to Greenpeace, and the judge had agreed with my contention that *Lophelia pertusa* could be harmed by drilling activities associated with oil exploration and production. However, the most significant aspect of the judgement was that the European Habitats Directive applied to deep-sea ecosystems along the continental margin of the United Kingdom and, by extension, to the rest of the European exclusive economic zone. These are the waters over which European coastal states have jurisdiction for the purposes of extracting resources. Apparently, the news of the judgement was such a shock that one of the opposing lawyers rushed off to the toilets to be sick.

In my mind, this had been a true David and Goliath situation. I had fully expected the wheels of government, the oil industry and 'the system' to have simply ground down Greenpeace and, by fair means or foul, defeated it in court or had the case thrown out. I felt the system was bound to be corrupt. Here was the starkest evidence that my cynicism was unfounded and the law had, despite the vested interest from government and the economic might of the oil companies, taken a fair view of the evidence and had upheld Greenpeace's case.

In the weeks that followed I was invited to a meeting between Lord Melchett, executive director of Greenpeace, and several representatives of government departments involved in the case. Melchett berated the sheepish civil servants for even contemplating fighting the judicial review that Greenpeace had brought to court. His reasoning was that it had been such a clear case

109

that he did not understand why they had contested Greenpeace's point of view. It was one of the most embarrassing meetings I have ever attended. The civil servants took their tongue-lashing like schoolboys humiliated by a headmaster for some wrongdoing. At the National Oceanography Centre there was little comment on what had come to pass, although my entire research group basked in the glow of the victory. There was a sense that the boss had done something good for the oceans.

In the coming months and years, the full impact of the judgement was to become clear. The oil companies still drilled west of Scotland and found oil. However, in 2001, Margaret Beckett, the Secretary of State for Environment, Food and Rural Affairs, declared the government's intention to set up a marine protected area around the Darwin Mounds. In 2002, the Mounds were closed to fishing under emergency measures, and after agreement with European fisheries ministers the area was finally fully protected in 2004. By 2013, 20 Special Areas of Conservation had been set up in the offshore waters of the United Kingdom, including the Anton Dohrn Seamount, parts of the Rockall Bank and the Wyville-Thomson Ridge. The Greenpeace case passed into European case law, and for the first time there was a mechanism to protect the biodiversity of Europe's deep seas. Scientists and legal experts studied the case, and I found I was often asked to participate in interviews or questionnaires related to these studies. Eighteen years later I was to hold a meeting with representatives of industry about a large report I was writing on deep-sea science for the twenty-first century for the European Marine Board. A scientist from one of the major oil companies was present at the meeting and the first thing mentioned was the Greenpeace case and my involvement. Clearly the defeat still rankled 18 years later. This is remarkable given that the *Deepwater Horizon* disaster had clearly demonstrated that oil accidents can seriously damage deep-sea life, including habitat-forming corals.

I revisited Rockall in May 2016 to investigate the coral habitats and seamounts in the area. This time, I went on the RRS *James Cook*, which was equipped with the remotely operated underwater vehicle *Isis*. The cruise – which was led by one of my ex-PhD students and attended by two of my postdocs and some representatives from English Nature – left from Southampton. We all stood on the monkey island watching the oil terminals and other shipping drift past as we sailed down the Solent. It was blue-water sailing down the Channel and then north. Past St Kilda and out towards Rockall, the coast of Scotland was rugged and atmospheric in mist and the golden light of dusk. On board we prepared our equipment, including microscopes and cameras for photographing specimens as well as the laboratories for the work ahead. The engineers prepared *Isis*, the ROV that had served us so well in the Antarctic. It was my first opportunity to see *Lophelia pertusa* reefs up close. I was not to be disappointed.

We dropped *Isis* onto the flank of the Anton Dohrn Seamount and began to move up the slope of the underwater volcano. ROVs always have to move up on a dive to avoid hanging the cable, which is attached to the ship, on some ledge or submarine cliff. Patches of coral were scattered around on boulders and exposed bedrock. The live coral varied from white to a pale golden colour, with yellow sponges and deep carmine-red soft corals, looking like balls with antennae sticking out from them. A large anglerfish stood its ground, a mottled grey-green and black with two golden-edged eyes, a huge sweep of mouth packed full of needle-sharp teeth, edged by a frill of protrusions that looked like a beard. Underneath this lurking predator a crab was sheltering, a very odd partnership none of us had seen before. As we moved further up the slope, grey sand and silt gave way to more and more patches of coral. Then there was more coral than sand, fields of dead coral with patches of live colonies growing on it, and *Lepidion*, the blue fish so commonly associated with *Lophelia*, cruising over the coral. Finally, we came upon the reef proper, a great bank of coral framework, much of it a dead grey-brown skeleton, covered in silt but growing on it a rind of white-and-golden *Lophelia pertusa* and the zigzag shape of the coral *Madrepora oculata*.

Tropical shallow-water and deep-sea cold-water reefs are mainly composed of frameworks of consolidated dead corals with a covering of live colonies, so this was a natural *Lophelia* reef as it should be. You can almost think of it as a tree formed of many individuals, the growing parts just under the bark, the leaves and the fruits, and mostly dead wood forming the bulk of the structure. I marvelled at the diversity of this underwater forest. Here and there black corals, perversely coloured orange or red, like great feathery bushes, grew. Pink, thick-spined pencil urchins were dotted around. *Lepidion* crouched in the framework for protection. There were also bright-yellow sea fans and brisingid sea stars, like elaborate upturned umbrellas of spiny arms, bright scarlet red in colour and sometimes in bunches of tens of individuals. Closer inspection revealed deep-red and bright-white anemones, blue sponges, filter-feeding sea cucumbers and myriad other animals. It was an underwater city as beautiful and stunning as any tropical coral reef, but certainly with fewer fish. The scientists watching the screens in the ROV control van murmured appreciatively among themselves as David, the ROV pilot, manoeuvred *Isis* over the underwater forest.

For me it was a special moment, and over the weeks to come we were to discover new areas of coral, some almost pristine but others nearly destroyed by human activity. Rockall treated us kindly, and we had many days of sailing over glass-calm seas in bright sunshine accompanied by dolphins and pods of pilot whales. That the chilly waters of the United Kingdom could offer such sights and harbour such treasures touched everybody: crew, scientists and engineers alike. It demonstrates that even in the relatively cool waters of northern Europe a rich diversity of marine life exists along our coasts and offshore in the deep sea. Because human knowledge declines with distance from the shore there is a temptation to believe that these deep, dark waters are lifeless and we can do what we want with little prospect of harm. I have heard both governments and corporations peddle this nonsense time and again over the last 30 years. Nothing could be further from the truth, and the more we discover about the ocean, the more we understand the importance of the life it contains to the maintenance of our planetary ecosystem, the one we depend on to survive. It is critically important that we invest in the science that will allow us to make rational and knowledge-based decisions on what we permit, as society, to happen in the ocean. The alternative is the current status quo where the services of the ocean, from which we all benefit now and in the future, are deliberately or accidentally destroyed for the enrichment of a few. The next chapter graphically exposes what happens when an industry exploiting the deep sea is allowed to develop without prior knowledge of the environment it is exploiting or the resource it is extracting.

4

# Deep-Sea Fishing:
## *Would You Clear-Cut a Forest to Catch the Deer?*

It was early in November 2011, and after a very pleasant couple of days unpacking equipment on the ship and investigating the local restaurants and bars, the RRS *James Cook* set sail from Cape Town. The sun shone down on us as the ship moved smoothly through the Waterfront harbour and out to sea. We all stood on the foredeck taking photographs of Table Mountain, which crowns the city majestically, and the ship swung to the east to take up the first station. Cape fur seals were lying in rafts dotted here and there and swam off lazily or dived suddenly as the vessel approached. Cape gulls drifted past and Cape gannets flew low over the water, passing by without paying heed to the ship.

As I stood looking out to sea, I contemplated our mission. Our aim was to investigate the seamounts of the Southwest

Indian Ridge, a range of submarine mountains stretching from the east of Madagascar to the Antarctic near Bouvet Island. The organisms living on the seamounts had never been investigated, but our models – which relied on information about the environments where corals had been previously sampled to predict where they might occur globally – suggested that corals were likely to be found on them and that there might be cold-water coral reefs present. Testing the models required exploring an area that had never before been visited by scientists to see if the corals were present. We also knew that the seamounts had been trawled for deep-water fish and I wanted to find out how extensive the damage had been. While deep-sea oil and gas production was a potential issue for deep-sea coral reefs on continental margins, fishing was a far more widespread activity, and bottom trawling had the potential to seriously damage seabed ecosystems. If I could measure the damage, I could prove how destructive this industry could be to the ocean if not regulated properly. And if I could prove it, perhaps I could help to make further changes in how the fishing industry operated in our seas.

After an hour and a half, we made a stop just outside the shipping lanes leading into Cape Town and deployed the CTD, a very similar instrument to the one we had used in the Antarctic, to sample water below the ship. Then it was time to calibrate the multi-frequency echo sounders, acoustic devices used to look at the behaviour of the animals living in the water under the ship. This was a very fiddly operation requiring the suspension of a small metal sphere beneath the echo sounders, which were mounted in the hull of the ship, using a combination of three lines deployed from what resembled fishing rods over the port and starboard sides. We had heard horror stories about how long this operation could take, a day or more, but we were very lucky with calm weather and immediately managed to get echoes from the sphere. While we were absorbed in this painstaking activity, Cape fur seals swam around the ship. They were very sociable, often gathering in small groups, prodding each other with their noses or scratching and grooming themselves with teeth or flippers. Occasionally, one would hang, tail up out of the water and head down below the surface. These are the largest of the fur seals, with males growing up to 360 kilograms in weight. Their population is reported to be stable at between 1.5 and 2 million adults but, controversially, 80,000 pups are clubbed and 6,000 adults shot a year in Namibia. This is justified on the basis of controlling the impacts of the seals on fisheries, but the skins are mainly exported to Turkey where a wealthy businessman controls most of the industry turning the pelts into fur clothing. Not only is this a significant animal welfare issue, but my view is that fisheries should be managed so there is sufficient food for wildlife, whether they are seals, seabirds, whales or other fish.

Later in the afternoon, our metal sphere and lines suddenly shot back towards the stern of the ship. The seals' playful curiosity had clearly got the better of them and they had come to investigate our device although, thankfully, they left it undamaged. We could see their movements on the echo sounders as they were blowing trails of bubbles while diving down to 70 or 80 metres below the ship, occasionally coming up chasing small fish. We had seen a large male trap some fish between the harbour wall and the ship while it was moored in Cape Town and then proceed to gorge itself for half an hour or more.

By two in the morning, after more than 12 hours of calibration, we recommenced our journey towards the first seamount, named Coral. The moon appeared between the clouds and the ship's wake glowed bright blue-green with bioluminescence, light produced by microscopic algae churned up by the passing of the ship. The Southern Cross shone in the night sky. It was a beautiful sight and I took it as a good omen for the next stage of our expedition.

Five days later we arrived at Coral Seamount, whose summit was 120 metres below the surface. We had mapped the seamount on a previous trip on the Norwegian ship *Fridtjof Nansen*, named after the famous explorer and oceanographer who made the first crossing of Greenland and a famous attempt on the North Pole. The narrow but fairly flat summit gave way to a gentle slope to 1,000 metres depth on its southern side, but less regular, steeper and craggy slopes to the north, east and west. The team were chattering nervously with excitement. The seamount had been named Coral because the deep-sea fishers had retrieved quantities of stony coral as by-catch in their trawls when they had explored it for commercially valuable fish stocks. This evidence indicated that if deep-sea coral reefs were anywhere it would be here. There was feverish activity as the ROV was prepared for immediate launch.

We were sailing with the *Kiel 6000*, a German ROV that had been brought in at short notice after our precious *Isis* had been sucked into the propellers of the *James Cook* on a trip to the Antarctic and was severely damaged. The *Kiel 6000* was launched over the stern of the ship rather than over the side like *Isis*, which was a more precarious operation involving a winch and cable through a large metal A-frame at the back of the ship and ropes to steady the swing of the vehicle. Once the ROV was in the sea we rushed into the control van. The seabed came into view and the ROV landed on a gentle slope surrounded by a bed of blackened finger-sized shells, which were the plates of long-dead stalked barnacles, subfossil remains, possibly thousands of years old.

After carrying out its scheduled checks, *Kiel 6000* lifted off and proceeded upslope, filming all the way. Outcroppings of black volcanic rock and boulders lay among the barnacle plates or scutes, studded with bright-purple, pink and white sea fans. There were also large, pink-red brittle stars with multi-branched arms called *Gorgonocephalus* because of their resemblance to the head of the monster Medusa from ancient Greek legend. The ROV began to cross sand, inhabited by the odd red crab. This in turn gave way to scatterings of dead branched stony coral. We began to get excited as coral fragments were generally the first sign of a cold-water coral reef. Sure enough, a barrier of coral loomed out of the darkness and the ROV lifted over it and began to traverse it. I was elated as our first dive pretty much confirmed the results of our computer models, which had suggested the presence of reef-forming corals on these seamounts.

The tangled framework of coral branches was mainly dead, as with a *Lophelia* reef, with just a few areas of bright-white skeleton with vivid orange-red living polyps. Large snails with conical, pearl-coloured shells could be seen feeding on these surviving areas. It was a cold-water reef formed mainly by a species of coral called *Solenosmilia variabilis*, known mainly from south of Australia and around New Zealand. I knew this was the first cold-water coral reef documented in the whole Indian Ocean, which, while being a sight we were hoping to see, still blew us away completely. Not only did the discovery of this reef mean we had achieved one of our primary research goals, but it was an incredible vision in itself, so festooned with life. There were bright-yellow, white and pink bottle brushes, beautiful types of sea fans. The coral framework was scattered with delicate lacework glass sponges that looked like the most fantastically intricate pieces of cut crystal. Here and there were clusters and mats of hundreds of bright-yellow, peach and pink anemone-like creatures. As the ROV passed over the coral, squat pink lobsters disappeared into holes. Close examination with the ROV cameras revealed pairs of eyes reflecting back the lights from inky black holes in the framework. It was like flying over a city and watching the residents flee to hide from an alien invasion. There were pink urchins with fine spines and bright-white spiny sea stars crawling over the reef. Occasionally, we encountered deep-purple fleshy sea fans with small flowerlike polyps and large red brittle stars clinging tenaciously to their branches. Large jellylike, brain-shaped, cream-white sponges could also be seen. It was sensory overload to try to take in the entire reef and its many inhabitants. I couldn't wipe the smile off my face at such an awesome sight – until we suddenly caught a glimpse of something that did just that.

# Deep-Sea Fishing

We had stumbled on a lost net. It had been dragged across the seabed scraping and breaking the coral behind it and wrapping up the broken chunks of reef in a tangled mess on the bottom. I felt a wave of cold fury wash over me. This was a seamount supposedly under voluntary protection by the deep-sea fishing industry. Because they had found corals in their trawls when exploring this seamount several fishing companies agreed to not fish it to prevent damage to the reefs indicated by the by-catch. Unfortunately, such voluntary measures only applied to a limited number of vessels from a few countries. The tangled net and ropes ran for a long distance along the seabed and eventually we abandoned trying to follow it, not least because of concerns about getting the ROV tangled in loose ropes and cable. The reason we were in this very remote part of the ocean was because we knew that the seamounts of the south-west Indian Ocean had been targeted by a deep-sea fishery. The fleet, which numbered dozens of vessels, had lasted for just two years before the catches plummeted, leaving just a few trawlers from New Zealand, the Cook Islands and Japan still fishing, specializing on trawling seamounts.

Although it seemed remarkable that fishermen might venture into such a remote part of the ocean, lying days from land between Africa and the Antarctic, red gold in the form of a deep-sea fish called the orange roughy had been discovered on these submarine peaks. Orange roughy was a highly profitable catch for the deep-water fishing industry, but here we were seeing first-hand the consequences of fishing using bottom trawls on a cold-water coral reef. Already we were gathering new information on the damage done to this remote range of underwater mountains before scientists had even had the opportunity to explore them and identify the fragile ecosystems living on them. It was my mission to collect this information, and we set about searching for more evidence of what had happened as though investigating a crime scene the size of an entire mountain range.

The orange roughy was a new name for a species of deep-sea fish originally called the 'slimehead'. They are found primarily along steep slopes of continental margins and on seamounts usually at 400 to 1,000 metres depth but can live deeper. Orange roughy are a bright red and rather rotund fish growing to up to 75 centimetres in length, although usually caught at a smaller size. They have a bony face with a star-shaped ridge of armour around the eye and a large mouth with a permanently sad expression. They have good reason for such a downcast look. Orange roughy were discovered by Soviet and New Zealand vessels over the Chatham Rise to the east of New Zealand in the late 1970s. This is a huge extension of the New Zealand continental shelf forming a plateau, including many seamounts and submarine hills out to 1,000 kilometres from the coast into the deep ocean. This zone of shallow water, reaching to within 350–400 metres of the surface, is the place where warm subtropical waters from the north meet the cold waters from the Southern Ocean. Such zones, where water masses of different temperatures, salinities and nutrient content meet, are known as fronts. Where the waters mix, the ocean is very productive. No doubt this was why there were such high concentrations of orange roughy and other fish over the plateau and before long catches reached 50,000 tonnes a year.

Obviously, the public were unlikely to warm to eating 'slime-head' so the marketers thought up the name of orange roughy. In addition to its unprepossessing name, the skin of the fish and layer of oil underneath causes diarrhoea if eaten and so these have to be removed in processing the fish, a technical problem that was easily solved by the fish processors. The oil has now found use as a beauty product for skin hydration. Orange roughy fillets have a very mild taste, they freeze well and are a good vehicle for a sauce, and the species soon gained popularity in New Zealand, Australia and the USA. Trawlers from Australia decided to try their luck on the seamounts to the south of the continent. South of Tasmania, they found their prey. One submarine feature called St Helen's Hill yielded over 17,000 tonnes of orange roughy in a single year.

This conical seamount or volcanic cone reaches 600 metres depth at the summit from around 1,000 metres at the base. The orange roughy spawned over the seamount forming dense shoals like a doughnut around the flanks. Catches were reputed to reach a tonne per second as the trawls were towed through the heaving masses of fish. Up to 78 fishing vessels fished the feature at one time; the trawlers queuing up to take a shot at the orange roughy aggregations. Fortunes were made by skippers and deckhands alike. At one stage, so large were the catches that the local fish processing facilities were overwhelmed and truckloads of orange roughy were dumped into landfill. Clearly this was not going to last, and the New Zealand and Australian stocks of orange roughy rapidly collapsed. This was not simply a result of intense fishing; it was because fisheries scientists failed to appreciate that the deep sea was very different in terms of productivity than shallow water. The fishing industry, in its search for new resources, partially driven by the overfishing and regulation of coastal and open ocean fish stocks, had started to exploit a world we knew very little about.

To understand the history and importance of the Indian Ocean mission I need to go back some years to the point where the collision between the deep-sea fishing industry and scientists like me first began. In 1992 I began my first postdoctoral position as a research fellow at the Marine Biological Association in Plymouth. I was becoming more and more interested in the deep sea because, of all the parts of the ocean I was familiar with, it harboured the most interesting and bizarre animals on Earth. In a world with no light, and under crushing pressure, life thrived and included creatures producing their own light, fish with enormous fangs and seemingly comprised entirely of stomach, and whole communities powered by chemical energy around deep-sea hydrothermal vents.

I was particularly fascinated by seamounts. These submarine peaks were very poorly studied but seemed to attract a wealth of life. As related in Chapter 1, I had first encountered them on an expedition to the Azores, a place where seamounts were associated with large numbers of fish, including predators such as sharks and tuna. I had also found photographs of the summits of seamounts with gardens of fanlike corals in a book called *The Face of the Deep* by Bruce Heezen and Charles Hollister.

It was during research for a paper I published in a science journal – a paper that identified for the first time that overfishing of seamount populations of fish and of corals for the jewellery industry was a significant conservation issue – that I discovered just how unusual the life history of the orange roughy was. Although as it turned out, this was not unusual for the deep sea. The fish that was appearing on plates across the wealthy countries of the world could be 150 years old or more. Orange roughy are very slow growing and do not mature until they are 30–40 years old. Compare these figures to Atlantic cod which live for up to about 25 years and mature after 2–4 years. In addition, female orange roughy do not spawn every year, unlike many shallow-water fish, but wait an unknown period until they have accumulated enough energy to produce their eggs. All of these traits have evolved in response to an environment where food is limited and rates of natural mortality low. To make matters worse, orange roughy gather in shoals over prominent features on the deep seabed to spawn, including seamounts. Modern echo sounders meant that these huge aggregations were easily detected from the surface and satellite navigation meant that fishers could easily find the same features again and again at the right time of year. So the slow rate at which the orange roughy reproduced combined with their ease of being fished, meant that fishery after fishery targeted at this species collapsed as the stocks were massively overfished. The fishers responded by searching for further unexploited seamounts, effectively mining the orange roughy stocks one after another.

Orange roughy was not the only species in trouble. The first industrial seamount fishery had developed in the late 1960s, more or less when I was born, on the seamounts of the mid-North

Pacific. Here the Japanese and Soviet fleets discovered vast shoals of fish spawning above the Hawaii-Emperor seamount chain. This range of 80 or more seamounts stretches over 5,800 kilometres from the Aleutian Trench to Loihi Seamount, to the south-east of Hawaii. So vast is it, that it even diverts the mighty Kuroshio Current flowing from the Philippines towards the Arctic. It has never been thoroughly explored even now, yet ocean-going trawlers began to reap an enormous harvest of a new fish with catches reaching 133,000 tonnes in 1969. The identity of this fish was not even known as they were being gutted and filleted on factory ships at sea. Not until 1972 was the fish identified as pelagic armourhead, a species that until then had been considered rare. Already, by that point, the catches had dropped to 20,000–30,000 tonnes and by 1977 had dropped further to 3,500 tonnes.

Little is known about pelagic armourhead. The fish may be semelparous, like Pacific salmon, spawning once over the Emperor Seamounts and northern Hawaiian Ridge before dying. The larvae live in the surface of the ocean and develop into sub-adult fish, which migrate to the edges of the Arctic where they feed for one to two and half years, accumulating great reserves of fat. The armourhead then return to the seamounts where they metabolize their fat reserves and feed on migrating layers of plankton and smaller swimming animals for three to four years while they mature. As with the orange roughy, pelagic armourhead gather above seamounts for spawning and it is here they were targeted by trawling fleets. Over the space of 10 years, nearly a million tonnes of pelagic armourhead were stripped off the Emperor-Hawaii seamount chain and the fishery collapsed. On the *Fridtjof Nansen*, the ship on which we first visited the Southwest Indian Ridge, we had caught a couple of these fish in a pelagic trawl, a net dragged through the water column rather than along the seabed. On the Southwest Indian Ridge, they were only found on a single seamount, Atlantis Bank, a sunken island with a flat sand-covered summit lying at 700 metres deep. They were magnificent-looking fish, up to half a metre long, with a deep blue-grey body, heavily scaled. They had a wedge-shaped armoured head with a large black eye ringed in silver, and a small mouth. A broad tail and spiny dorsal fin ran down their back. Old Russian records indicated that Atlantis Bank was the only seamount where these fish were caught in any numbers on the Southwest Indian Ridge and, as mysterious as it is, it would seem that only this seamount provided the right habitat for this species in this part of the Indian Ocean.

Nowhere in the fisheries literature did I find any reference to the impacts on the wider ecosystem of removing such a large biomass of fish. Large numbers of pelagic armourhead were found in the stomachs of sei whales hunted by the Japanese whaling fleet. Presumably this source of food for the whales was simply removed in the North Pacific by this rapacious fishery. Likewise, with orange roughy, there was no indication of the wider implications of the removal of multiple stocks from seamounts. There was another more disturbing problem with these fisheries, though.

In 1997, after working at the Marine Biological Association in Plymouth, I had taken up a fellowship at Southampton, and while there a scientist from Australia, Tony Koslow, passed through to give a seminar on seamounts. Tony was excited to find the author of the 1994 seamount paper at the Oceanography Centre as he did not realize I had moved to Southampton. What he revealed in his talk that day remains with me as though I were a witness to some great natural disaster or horrible accident. Tony discussed seamount fisheries and the energetic basis for how seamounts could host dense populations of predatory fish in the deep sea, an environment that we knew was food-limited. I had gathered evidence during the research for my paper that a likely mechanism was the trapping of migrating zooplankton and small swimming animals (micronekton) by the seamount. It is a little known fact that the world's largest migration happens every day across the ocean.

At dusk millions of zooplankton and micronekton, including jellyfish, shrimp, small fish and squid, migrate from deep waters to the surface to feed. The darkness of the night hides them from the fast-swimming, agile predators of the sunlit waters of the shallow ocean. Some fish migrate as much as 1,600 metres to reach the surface. At dawn these animals flee back into the safety of the dim waters of the twilight zone, or even deeper into the bathyal zone below 1,000 metres depth. Overnight layers of these animals drift over seamounts and as they dive at dawn to reach safety, their migration is blocked by the shallow seabed of the seamount summit and flanks. Fish and other animals on the seamount attack the trapped migrators desperately trying to reach the deep sea in waters that rapidly become illuminated by the rising sun.

On the *Fridtjof Nansen* we had actually observed this

phenomenon on a multi-frequency echo sounder over Atlantis Bank. This device sends pings of sound, which are reflected from animals in the waters underneath the ship. The reflections were displayed as coloured pixels on the screen moving from blue to red as increasingly dense reflections were detected. A dense blue layer could be seen sinking towards the summit of the seamount and then dense red patches rose from the seamount and intercepted the migrators. These were the seamount predators, and shooting the net through these shoals retrieved pelagic armourhead, another commercial species, the alfonsino, and bigger predators including the scabbard fish, a silver fish shaped like a straight sword, growing to over a metre long with a jaw packed with needle-sharp teeth. These fish were found to have stomachs full mainly of lantern fish and squid typical of the migrating layers.

During this cruise we observed on the echo sounder that the seamount predators responded to anything in the water. When we lowered instruments into the waters above the seamount we could see the fish diving for the safety of the seabed. It is for this reason that fish like orange roughy and armourhead are often fished using bottom trawls.

In his seminar, Tony described the seamount fisheries south of Tasmania. Robust gear had been developed by the New Zealand deep-water trawlers to fish on the rocky rugged ground of seamounts. The trawls were equipped with heavy steel doors, weighing several tonnes, to keep the nets open while being dragged along the seabed. They also had rollers at the bottom of the mouth of the trawl to enable it to roll or hop over rocks.

# The Deep

Tony showed a slide of an unfished seamount at a depth of about 1,300 metres covered in a rich coral reef. Urchins could be seen crawling over the coral framework which was scattered with delicate glass sponges and yellow crinoids, much like the sights we would come to see on Coral Seamount in 2011. Then he showed one of the fished seamounts. I can only describe it as looking like the aftermath of a nuclear bomb explosion. The seabed was scraped completely clean, leaving behind a fine pale dust and gravel. The marks of trawl doors scarred the scene, running from the bottom of the picture into the darkness beyond the lights of the camera. A single red soft coral was left growing in the foreground. There was a sharp intake of breath and murmuring from the audience. The scene of utter devastation left me feeling sick. Something had to be done.

In the early 2000s two significant efforts came together which were to prove important in shifting management practices in deep-sea fisheries. The first was the Census of Marine Life. This was an international effort sponsored by the Alfred P. Sloan Foundation and driven by the late Fred Grassle, a passionate American scientist who had undertaken the first biological investigations of hydrothermal vents near the Galapagos Islands. The aim was to explain the distribution and diversity of life in the oceans. Included within the programme was the Census of Seamounts, led by Malcolm Clark, a deep-sea ecologist and fisheries biologist from New Zealand. Malcolm was a tall bearded man who had already been dealing with the deep-sea fishing industry in New Zealand for several years. It had been New Zealand trawler captains who had developed the equipment and techniques to fish for orange roughy, first in the waters of New Zealand and then beyond in the high seas. This was the zone of the ocean beyond the jurisdiction of coastal states, usually set at 200 nautical miles from the coastline. Malcolm worked for the National Institute for Water and Atmospheric Research (NIWA) on the management of deep-sea fisheries. He was incredibly hard-working and under his leadership scientists around the world began to coalesce what was known about the ecology and the impacts of deep-sea fishing on seamounts. At the same time, I began to advise the International Union for Conservation of Nature (IUCN) on issues related to the conservation of deep-sea ecosystems in the high seas.

The IUCN had an unusual political position in being an intergovernmental organization. This gave it status as an observer in political meetings at international level, including at the UN General Assembly and the UN Food and Agricultural Organization (FAO). The former had some oversight of how the oceans, especially the high seas, were managed. The latter was supposed to coordinate actions to manage fisheries, including those of the regional fisheries management organizations (RFMOs), which implemented fisheries management regulations on the high seas. Under the leadership of Carl Gustav Lundin, head of the marine programme, IUCN was developing the case that the high seas, as well as coastal ecosystems, also needed protection. Carl had previously worked for the World Bank and was well equipped with the diplomatic skills required to understand the political landscape of ocean governance. Also working with the IUCN was Kristina Gjerde, an American lawyer dedicated to improving governance of the ocean, especially with respect to deep-sea fisheries. I also encountered Matt Gianni, an ex-fisherman who had worked on a deep-sea fishing vessel off the Pacific coast of the USA. He had been so appalled by the destruction he had seen of deep-sea ecosystems that he had joined Greenpeace as an ocean campaigner. Matt worked tirelessly on the whole deep-sea fishing issue and had great knowledge of what was going on at the various RFMOs in charge of deep-sea fisheries on the high seas.

The fishing industry was in denial. As photographs of the devastation caused by deep-sea bottom-trawl fishing began to come to light there were efforts to suggest that these were isolated cases. However, as the photographs began to accumulate, evidence of the damage done by bottom-trawl fisheries in different parts of the world became undeniable. Off the coast of Norway, a scientist produced video evidence of the destruction of *Lophelia pertusa* reefs by trawling. Surveys along the edge of the continental shelf revealed smashed and scattered fragments of coral, turned over boulders, deep furrows from trawl doors and various items of lost fishing gear including anchors, gill nets, trawl nets and rope. The rock hopper trawls had allowed fishers to drag their nets over rough ground with the result that an estimated 30%–50% of the coral areas off Norway had been damaged or destroyed.

Ironically, it had been the fishers who had called in the scientists because of concern that falling catches of fish were resulting from loss of the habitat represented by the coral. The cruises by Southampton Oceanography Centre over the Darwin Mounds had obtained acoustic images of mounds obliterated entirely by trawl tracks running through them. In the north-west Atlantic, video captured from the Corner Rise seamounts showed summits scraped clean of life and with the clear marks of trawls and trawl doors all too evident on the bare rock left behind. So heavy was the impact of trawling that it had even broken up the hard manganese crusts on the surface of the seamount. These seamounts had been fished by Soviet trawlers from the mid-1970s to the mid-1990s. Greenpeace became involved and followed deep-sea trawlers out to their fishing grounds in the South Pacific. I found the images they gathered of huge red sea fans with stems as thick as tree trunks and larger than a person, even with most of the branches broken off by the trawls, being dumped back in the ocean particularly upsetting. Evidence was now coming through from isotopic studies of coral skeletons that some species of black coral could reach more than 4,000 years in age. These were the bristlecone pines of the deep. How old these huge specimens were that were being tossed back in the ocean was a matter of speculation, but I was sure they must have dated thousands of years. Likewise, reefs formed by *Lophelia pertusa* were also discovered to be thousands of years old.

The evidence revealing the true extent of the damage the fishing industry was doing to our oceans, destroying these ancient ecosystems seemingly without a grasp of, let alone a care for, the consequences, was deeply concerning. In 2002, in the journal *Nature*, Daniel Pauly, a world-renowned fisheries scientist drew the analogy that trawling was like clear-cutting forests in the course of hunting deer. The picture this conjured of the destruction of complex ecosystems of the deep sea, causing habitat loss for thousands of other species, was extremely aptly put. Subsequently scientists, environmental non-governmental organizations (NGOs) and advocates replaced 'deer' with 'squirrels' reflecting the even smaller catches from deep-sea trawling. The stage for the battle between the fishing industry and those scientists who understood the danger the ocean was in at their hands was set. Beginning in 2003 at the World Parks Congress in Durban, South Africa, clashes between scientists and conservationists versus the fishing industry, their lobbyists and fisheries managers began. At the invitation of IUCN I gave a talk on the deep sea, which included the latest findings on the rich diversity associated with seamounts and cold-water coral reefs and the threats posed to these ecosystems by fishing. Here I encountered Javier Garat for the first time, a lobbyist for the Spanish fishing industry. Javier was always immaculately turned out, in a blazer, smart tie and chinos, and was a passionate defender of the Spanish fishing industry. He was born into a family that founded an industrial fishing company, Albacora, trained as a lawyer and worked for Cepesca, a fishing lobby organization. For his efforts on behalf of the Spanish fishing industry he had been honoured by being made a Knight of the International Order of the Golden Fleece.

Our first clash occurred in a small room designated for side events. It was in front of a panel discussing the management of fisheries on the high seas. At one point Javier stood up and eloquently stated that deep-sea fishing avoided corals because they were dangerous to the fishing gear and that most took place on soft bottoms – mud – where there was no life. My alarm bells immediately rang. The absence or near absence of life in the deep sea was the same fallacy that Shell and the UK government had put forward during their dispute with Greenpeace over the *Brent Spar* abandoned oil platform. I immediately intervened from the floor and stated that deep-sea sediments had, in fact, been found to harbour a very high diversity of marine life and that this reached a maximum at mid-slope depths, somewhere between 1,000 and 3,000 metres. Javier visibly faltered and then back-pedalled saying that he had not meant there was *no* life in such ecosystems. To me it was a hint of what may be being fed to politicians by the industry when there were no scientists in the room to qualify facts about deep-sea ecosystems affected by fishing. That key decisions were potentially being made based on skewed or false information was deeply concerning. That scientists weren't even being given the floor on such matters made me burn with frustration. As I was to discover later, Cepesca and other fishing industry lobby organizations were more or less permanently present at the European Parliament arguing for less stringent quotas and more taxpayers' money in the form of subsidies to support modernization of fishing fleets with better vessels and more efficient fishing gear. How were we going to make sure our voices were heard and that politicians had all the information in front of them when making such crucial decisions about what happens to our seas?

There followed a whirlwind of meetings, the Conference of Parties of the Convention of Biological Diversity in Kuala Lumpur, Malaysia, and the World Conservation Congress in Bangkok, Thailand, both in 2004, and then the FAO Fisheries Committee meeting in Rome in 2005. The American lawyer Kristina

# The Deep

Gjerde, Carl Gustav Lundin and Matt Gianni, all working with
the IUCN's marine programme, were ever-present, either trying
to raise understanding of the biodiversity and vulnerability of
deep-sea ecosystems or arguing for better governance of the high
seas – the waters beyond national jurisdiction. Out in these waters
there was no legal framework for the protection of diverse and
fragile ecosystems like cold-water coral reefs or seamounts. This
was a historic legacy in that when the United Nations Convention
on the Law of the Sea (UNCLOS) was negotiated, there was a
low level of awareness of the diversity of life in the deep ocean.
Additionally, the concept of 'freedom of the seas' meant that
states resisted attempts to regulate activities beyond their waters.
Across vast areas of the ocean there was no oversight of fishing
activities, which were supposed to be managed by the flag state
of whatever vessel was out there. At best, fisheries were run by
RFMOs, closed shops formed of representatives of the governments of flag states, which met and agreed fishery regulations
and quotas by consensus. These organizations still operated with a
low level of transparency and the decision-making processes were
generally driven by the lowest common denominator, namely
what the most aggressively exploitative fishing nations would agree
to. Lobbyists were ubiquitous and although these organizations
have scientific committees, much of the focus was on setting sustainable quotas for target fish species, not on the environmental
impacts of the fishing. Lack of evidence was often used by states
driven by fishing industry interests, such as Japan, to stifle action
aimed at lowering quotas or modifying fishing practices to protect
the environment.

Of course, some RFMOs were better than others, where the responsible states invested in more scientific advice and the objectives of management were based not only on sustaining fisheries, but also the environment. On the whole, however, what I saw was very worrying. It seemed we had not learned much from the experience of the International Whaling Commission where states presided over the successive decimation of whale populations moving from the largest, most valuable species, to the smallest. Arguments over quotas by states like Japan strangled any hope of bringing the numbers of whales killed under control, and this was compounded by massive under-reporting of catches by Russia (then the Soviet Union). Only when there was public outcry, following the revelations by Greenpeace of what was unfolding, was this slaughter stopped, otherwise many species of whales would likely have vanished from the ocean.

Although many of us live in democracies, it is still very difficult to bring change around an issue involving environmental damage by an industry when other economic and social issues place increasing demands on politicians' attention. This is even more the case with an issue like deep-sea trawling where the activity is taking place in distant waters beyond the public's perception. Making the space for the voice of civil society to be heard in national and international governmental bodies is incredibly difficult and demands a variety of skills and strategies, ranging from public education and publicity, to personal contact and discussion with individual policymakers, and intimate knowledge and understanding of, in this case, environmental law. Experienced communicators are therefore a key part of any campaign to change public policy. It was during the flurry of meetings in 2004 that I first ran into Mirella von Lindenfells and Sophie Hulme, co-founders of Communications INC, a company that specialized in working with NGOs on communicating issues to the public and policymakers. Mirella was well-spoken and very vocal. She was passionate about the ocean and issues related to its abuse. Sophie was a serious and hard-working campaigner. We all immediately struck a chord. Sophie had worked for NGOs, such as Amnesty International, on very different issues, and Mirella's background was at Greenpeace. However, we all had a shared view of a better future where both people and the environment were treasured and treated with more consideration. As part of the work informing politicians about the devastation caused by deep-sea fishing, Communications INC organized a scientists' tour of Europe so that experts like me could speak to government representatives face-to-face about the issues associated with deep-sea bottom fishing. On one such trip to Poland, our journey finished with a presentation and discussion with a Polish fishing minister whom we were trying to convince to support regulations in Europe to fish more sustainably and who looked quite puzzled for much of the meeting. 'But we do not have this type of fishing I think.'

'Poland might not, but other countries in Europe do, and your views and your votes count in the European Parliament,' Sophie patiently explained.

Every member of the European Parliament and Commission has influence on decisions made with respect to the management of fisheries undertaken by the European fishing fleet. Europe is a major player in deep-sea fishing globally and so informing European politicians of the devastation caused by deep-sea fishing and winning their support for reform was critical. There followed an event at the European Parliament arranged by Carl Gustav Lundin, where I presented on the biodiversity of the deep sea and the impacts of bottom trawling on seamounts and cold-water coral ecosystems. Javier Garat presented the view from the fishing industry forcefully. He had brought along a Spanish fisherman who tried to put across how impoverished the industry was. During the question and answer session, one of the government representatives of a European state claimed that Irish trawlers had deliberately destroyed an orange roughy stock off of Ireland simply to prevent the French and others from 'taking their fish'. If true it was another example of treating living populations of fish as mining resources. I hope that the audience of European commissioners and parliamentarians, sceptical at first, was swayed by the evidence of overfishing and environmental damage that was laid before them. That some of their own members were seeing the results of careless overexploitation and destruction of stocks added even more weight to the arguments we presented. This was critical because the European Union was a strong voice within the UN General Assembly.

Gradually, the tide at the UN in New York began to turn with a series of General Assembly resolutions. Beginning in 2004, these called for a cessation on destructive fishing practices that damaged vulnerable marine ecosystems, including seamounts and cold-water corals. This was reiterated in 2006 with another UN General Assembly call to put in place management measures to prevent damage to the aforementioned deep-sea ecosystems. Here we began to run into elements of the fishing industry involved in the South-West Indian Ocean. Representatives of industry, some members of the UN Food and Agricultural Organization and the RFMOs knew each other very well. I had seen them dine together at meetings, drink together into the evenings and they were vociferous in their defence of an industry that had engaged in the indefensible destruction of the marine environment. As the conservation arguments gained ground, discussions at meetings from the fisheries industry became less logical and more difficult to follow. Many of the scientists who produced evidence of the damage fishing was doing in the deep sea were verbally lambasted for poor science or extrapolation of results based on local studies. I personally witnessed this happen when the scientists in question were not in the room to defend themselves and this was a strategy clearly aimed at undermining their credibility in the eyes of the audience, usually politicians. It was an aspect of the battle that I found particularly underhanded. No doubt I came in for similar treatment when I wasn't around to defend my work. Some of the representatives from the UN Food and Agriculture Organization, based in Rome, and some of the deep-sea fishing companies knew the writing was on the wall at the United Nations and had worked together to declare voluntary protected areas along the Southwest and Southeast Indian Ridges to protect cold-water coral reefs and areas of special scientific interest. It was a step forward, and Coral Seamount was one of the areas under voluntary protection.

Despite the UN General Assembly resolutions, actions to reduce the damaging effects of deep-sea bottom trawling were very slow or insufficient. However, by 2007 sufficient pressure had built up for the UN General Assembly to act decisively. The UN FAO was directed to take action to improve the environmental sustainability of deep-sea bottom fisheries in the open ocean. I was invited to participate in a workshop in 2007 to help develop new guidelines for the management of deep-sea fisheries on the high seas. The meeting was held in a hotel in Bangkok. As well as Matt Gianni and Kristina Gjerde, there were a number of scientists, all experts in deep-sea fisheries or seamount ecosystems, many of them colleagues who I had worked with previously. Representing the industry was Graham Patchell from the fishing company Sealord, owned by New Zealand (Māori) and Japanese interests, and there were representatives from the FAO and various fisheries bodies, including Ross Shotton who had been involved in setting up the voluntary protected areas in the southern Indian Ocean. The discussions were generally constructive and business-like. There was a realization among all parties that the UN, and therefore the FAO, were under significant pressure to improve the management of deep-sea bottom fisheries.

The weight of scientific expertise around the table meant that the discussions were evidence based. A particularly sensitive issue was around what was meant by temporary damage to deep-sea ecosystems. If fishing was impacting ecosystems but they were recovering before further impacts, then such damage was deemed acceptable in the course of catching fish. A robust discussion took place about whether a time limit for 'temporary' impacts should be established. This was critical to ensure protection of cold-water coral reefs and other fragile communities of animals.

If a time limit was not established, then fisheries managers would be at liberty to impose their own definition at the risk that fragile ecosystems with a limited ability to recover would never be protected. I pushed very strongly that what was meant by temporary should be defined and in the end we agreed that this meant a lack of recovery over 5 to 20 years. It was a particularly gratifying moment for me and I felt as though I had been able to use my expertise to influence the guidelines on management of deep-sea fisheries that were taking shape. Many of the scientists around the table had witnessed destruction of deep-sea ecosystems that were likely to take hundreds of years for recovery if they ever recovered at all.

The IUCN asked me to attend the meeting at the UN FAO headquarters in Rome in 2008 where state representatives agreed on the final text for the guidelines. I spent several days watching with fellow scientists as lawyers and representatives of fishing ministries argued over the final wording of the guidelines, even down to whether the words 'and' or 'or' should be placed in one area of the text. The sad thing was that such tiny differences in wording made a significant impact on the implementation of the guidelines out in the ocean. When the final text was agreed we collectively breathed a sigh of relief. After five years of effort by scientists, NGOs and the IUCN, we arrived at the first serious international effort to deal with the horrific destruction caused by bottom trawling on vulnerable deep-sea ecosystems.

For me there was a real feeling of satisfaction at a job well done in the face of open opposition from the industry and an insidious relationship between organizations meant to be implementing sustainable fishing on the high seas and vested business interests. Flag states and regional fisheries management organizations now faced the task of implementing the new guidelines, including establishing sustainable catch levels for deep-sea fish, protecting vulnerable marine ecosystems through fisheries closures and developing protocols to identify new areas that were characterized by high by-catch of corals and other fragile marine life. This was not the end of the matter, however. As I set out on my first expedition to the Southwest Indian Ridge in late 2009 and then to the Antarctic in early 2010, I was writing a report on the implementation of the FAO guidelines with Matt Gianni. Adherence to the FAO guidelines was very uneven across states and RFMOs. The report Matt and I authored concluded that while some countries and RFMOs were taking the guidelines seriously others were not. The UN General Assembly pushed for better implementation of the guidelines through further resolutions in 2010 and 2012. However, it had little ability to force states to follow the rules, and in large parts of the ocean it was still the Wild West.

Back in the Southern Indian Ocean on the RRS *James Cook* in 2011, we progressed from seamount to seamount, seeing more and more evidence of fishing. We dived Melville Bank, a very complex seamount with two distinct peaks with a jumble of cliffs and steep terrain. The slopes were covered with broken black basalt and there were small pink-orange lobsters, *Projasus*, scattered here and there. We came across slopes of fragmented coral and sediment and then a rope looming out of the darkness stretching above the ROV. Attached to it was a single lobster pot, not altogether different from the kind my grandfather used in Ireland. That lobster fishing was going on out here, five or six days of sailing from the nearest land, was just inconceivable. We backed away from the rope in case we had inadvertently gone under further fishing gear. We traversed the seamount further, encountering ridges of bedrock with bright scarlet-red and purple sea fans growing where they were most exposed to the current. On steeper terrain we saw fields of bright yellow sea fans.

Volcanic cliffs were home to slender grey cardinal fish, another species targeted by seamount fisheries that could live to 70 years or more. But the evidence of fishing was ever-present. On one cliff we found two tows of lobster pots tangled around each other. One tow was of the traditional net-covered pots of the type familiar from Ireland, whereas the other was formed of pale-blue plastic pots which resembled the cages we used to take the cat to the vet. There were dozens of pots. I realized this was a completely unknown fishery, undeclared and unregulated yet going on out in the distant Indian Ocean. It was yet more evidence of the complete lawlessness of the high seas and what went on out of sight of the world. Then we spotted what appeared to be an engine gasket, obviously thrown over the side of a fishing vessel.

The seamount summit which came up to within 100 metres of the surface was formed of a series of black basalt ridges festooned with growth. Small, feathery grey-black coral colonies gave the seabed a furry appearance and in among these were colonies of bright-golden, yellow-orange *Balanophyllia*-type corals. These resembled clusters of flowerlike anemones like miniature suns. There were also bright-white stylasterids or hydrocorals, like slender antlers growing from the seabed. In one ravine there lay an anchor, probably part of a net or line left behind on the seamount. It was encrusted with life, and a rope attached to it trailed away into the darkness. As we moved up towards the summit of the seamount we were pursued by a group of wreck-fish. These were huge slate-grey, broad-bodied fish, with large golden eyes and a great predatory maw slung low on the head. They were hunting in the lights of the submersible but were too slow. Large silver jacks streaked past the ROV to snap up any unwitting smaller fish caught in the lights of *Kiel 6000*. The amount of life on Melville Bank was breathtaking. Michelle, the postdoctoral scientist working on the seamount project, suddenly called out: 'Stop! Stop!'

We all peered at the video screen, at a small shape hovering above the seabed. It resolved itself into a box-shaped pale fish covered in black spots.

'Oh my goodness, that's *Ostracion cubicus*. I've seen them on the coral reefs in Madagascar!'

The rather dramatic announcement caused some chuckles. It was indeed the small coral reef puffer fish and we all marvelled at what it was doing so far south and at 200 metres or more depth below the summit of a seamount. Michelle knew the animal well as she had run an NGO called Reef Doctor in Madagascar where the fish was common in coastal waters. The camera zoomed in and we could see the pectoral fins working like crazy, propelling the clumsy-looking animal along the seabed. Its black eye glittered with blue-green sparkles. It wasn't the only reef visitor.

# The Deep

Near the summit we saw rainbow-coloured wrasse scooting from hiding place to hiding place. This was evidence to me of an old idea that seamounts might act as stepping stones for animals to cross large distances of the ocean over time. The seamounts were like islands of shallow water lying in the wide deep sea. We also stumbled across a huge spiny lobster, pink with red-and-white-banded legs and crouched under a ledge. This was what the lobster pots were all about. Indian Ocean spiny lobsters were a major target for fisheries off South Africa and the Crozet Islands in the sub-Antarctic to the south. We only saw a few of these spectacular animals on our dives at Melville Bank. It was likely many had been taken by the fishery. As we reached the peak of the seamount it was like stepping into a busy street in London – if London was a giant aquarium. Clouds of greenish striped morwongs hovered in the dark. Wreckfish cruised past the ROV eyeing the vehicle with what appeared to be intense curiosity. There were not only the slate-grey ones but also a darker species with pale patches. These were Hapuku wreckfish. A slender grey-blue shark cruised past lazily stroking its tail through the current. Stretching out into the darkness was the field of gold, white and grey corals, with smaller fish at their business, swimming to and fro, in and out of nooks and crannies.

On our final dive on Melville Bank *Kiel 6000* tracked along one of the steep slopes of the seamount. We crossed fields of broken black boulders, sand shoots and ridges. Growing on these were huge bright-white stalks crowned by an intricate goblet-shaped sponge. We also found a strange red jellyfish attached to the rocks with fine filaments. We stopped to try and sample it, but the animal released itself and floated off in the currents. Then, in the distance, we spotted a bright-red fish. As the ROV approached it orientated towards us and I saw that it was an orange roughy. Here was the fish that we had all seen so often in photographs on the decks of fishing boats. Those animals were a sickly orange colour with ragged tattered fins, mauled by the trawl net as they were dragged to the surface in the thousands from their amorous activities in the deep and dark. Seen in its natural habitat, it was a beautiful animal. A deep pinkish-red with rounded fins spread about its body like a butterfly's wings and a fine forked tail. The dark eye stared at us from the skeletal bony plates of the head and a downturned mouth gave it an eternal expression of sadness. It drifted away into the inky blue-black leaving us all happy that there were at least some of these majestic fish left on these seamounts. I was so moved that I wrote a story about it for my children.

Further along the ridge, on a seamount called Sapmer Bank, we found a fine-meshed gill net stretched across part of the seamount. Trapped in the net and suspended above the seabed was one of the large spiny lobsters weakly struggling against it as it slowly died. Even the ROV pilots wanted to risk their multimillion-euro robot to rescue the animal, so distressing was the sight. We followed the net for a short distance, but it continued to stretch away into the darkness, dead fish lying below it attracting yet more animals to be snared. This was the first real example of a ghost net I had witnessed in the deep sea: fishing gear lost on the seabed but continuing to fish hungrily, trapping all comers whether they be fish, crabs or lobsters. What made matters even worse was to see a Japanese trawler turn up, shoot its nets and proceed to track along the edges of the seamount. We were obviously in the way.

The final seamount of the cruise was Atlantis Bank. This seamount had not originated as a volcano but as an upthrust piece of the Earth's crust. Atlantis was therefore of scientific importance because it comprised deep rock of importance to geological study and the understanding of processes in the Earth's crust and mantle beneath. It had previously been drilled by the International Ocean Discovery Programme, a large project aimed at understanding the Earth's history through studying marine rocks and sediments. Once it had been an island, and wave action had worn its summit into a flat plain encrusted in limestone with a thin drape of sand. Small sharks and the odd *Projasus* lobster were dotted around and a strange white-and-red fish with a long snout, a bellows fish, could be seen hovering just off the seabed. These fish resembled a multicoloured old-fashioned set of bellows and were related to seahorses. Here and there were jagged outcroppings of black rock, some of which were home to great trees of white corals festooned in turn by sea urchins with particularly thin, sharp spines called *Dermechinus horridus*. They were certainly horrible to try to pick up when sampled by the ROV. Towards the edges of the seamount there were more interesting areas. We found colonies of bubblegum corals, several metres high. Tightly curled around their branches were large brittle stars and other animals, such as feather stars and gorgon's head basket stars. There were fields of red anemones being attacked by large red sea spiders from the genus *Colossendeis*, the size of dinner plates. There were also delicate Venus basket sponges, animals that appeared to be spun out of the finest embroidery of silica, forming clusters of slender tubes. Inside each one was a single pair of shrimps, one male and one female, a new species. Here and there we saw a pelagic armourhead paddling over the seamount. Then on one flank of Atlantis we were rewarded with the final glimpse of an orange roughy. This one was pale pink and sat hovering with its nose touching the bare rock of the seamount. As we approached, the animal slowly turned white before scuttling off into the distance. Orange roughy were clearly not built for speed, and the fish waddled from side to side as it swam off into the darkness.

The struggle to manage what happens out in the furthest reaches and depths of the ocean continues today. The European Union finally agreed to limit fisheries for deep-water species to a maximum depth of 800 metres after years of campaigning by the Deep Sea Conservation Coalition and other NGOs. The limit should have been shallower, at 400 metres, as it is at this depth that the catch of slow-growing deep-sea species increases dramatically relative to the catch of commercially valuable species. Still the UN General Assembly is asking states and RFMOs to improve implementation of the FAO guidelines. In some parts of the world there has been real progress. Areas in the North Atlantic, South Atlantic and Antarctica where vulnerable marine ecosystems are present have been closed to fishing. Populations of orange roughy in some areas, particularly those characterized by high productivity, such as the Chatham Rise in New Zealand, have shown signs of recovery.

However, in other areas, such as the North Pacific, there is still little idea of what is being caught and where and certainly no attempt to moderate environmental impacts.

In the southern Indian Ocean there is now a new regional fisheries management organization and some of the seamounts we investigated have been protected from bottom fishing. Progress is, unfortunately, very slow. As I related in the previous chapter, in 2016 I was again on the RRS *James Cook*, this time investigating the banks and seamounts lying to the west of Scotland together with colleagues from the University of Plymouth and the Joint Nature Conservation Committee. Here we did, to our delight, find areas of spectacular coral growth now protected from bottom trawling. However, devastatingly, we also found areas where reefs were gradually being smashed out of existence, the trawl track clearly visible running straight into the coral. In deeper waters over a seabed of muddy sand, areas that were trawled were more or less empty of life compared to those which had not been fished. The odd sea cucumber remained and many of those were visibly injured.

Seamounts are rich biological oases in the ocean. They not only harbour rich coral communities and a wealth of species in themselves, but are also important feeding areas for many species of ocean predator, including large fish, such as tuna and sharks, as well as seals, whales and seabirds. They are also important breeding sites for certain fish and even act as navigational waypoints for animals such as turtles that undergo transoceanic migrations as part of their lifecycle. The destruction of fragile seamount communities and other vulnerable marine ecosystems has been rapid and largely undocumented, what scientists have recorded is the tip of the iceberg. This destruction continues today, and added to it are threats from new industries in the ocean, such as deep-sea mining to which I'll return later. Protection of such deep-sea ecosystems, especially those lying in the high seas, waters beyond national jurisdiction, is a matter of utmost urgency. Yet, as extraordinary as it may seem, there is no legal framework for protection of biodiversity beyond national waters, the largest part of the Earth's biggest ecosystem. Since 2004 an informal working group under the auspices of the UN has been working on how to protect biodiversity in the high seas. In 2017, both their work and pressure from scientists, NGOs, governments and more enlightened politicians who recognize the importance of the ocean, finally drove states to enter into negotiations to develop a new internationally binding legal instrument to protect marine biodiversity beyond coastal waters. In 2018 the UN General Assembly held the first intergovernmental conference to discuss these new international regulations and how to implement them. Further conferences are planned in 2019. Some states are opposed to such regulations or are trying to ensure they are weak. The decline of international relations as a result of the rise of nationalism is not helping. Fortunately, more progressive nations recognize that an urgent global effort is needed so that the biodiversity of the ocean can be maintained along with the many critical ecosystem functions it performs for us and for future generations. With many states looking towards their ocean as a new area for economic growth, including the deep sea, the stakes are higher than ever.

5

# Reef Encounter:

*How Do We Prevent the Destruction*
*of the Ocean's Most Iconic Ecosystem?*

It was the summer of 2015 and I was on my way to the Caribbean to join the Thinking Deep expedition, which was aimed at looking at the connections between the shallow parts of coral reefs and the zones below the normal depths to which scuba divers swim. Shallow-water reef-forming corals require light to survive but can actually live much deeper than most people realize: the record is 165 metres, which is more than three vertical lengths of an Olympic swimming pool! Some scientists had suggested that these deeper areas of reef might be important in the survival of coral reefs into the future, and this was a question we had set out to resolve in the Bay Islands of Honduras.

As with many tropical islands the world over, getting to the expedition was an adventure in itself. I had flown from London to Miami, where I stayed overnight, then on to Roatan, one of the larger islands in the Bay Islands group. After an anxious wait in the small airport and having been passed from one uniformed person to another, I finally managed to get on a small twin-engined Cessna. It was full of tourists headed to Utila, the neighbour to Roatan, a destination with the reputation for having the cheapest scuba diving instruction in the world. Despite the cramped interior of the aircraft I had a good view of the island as we took off, which still appeared to be covered in lush forest for the most part but with development along the coastal lagoons very apparent. Once we had landed, I took a tuk-tuk into Utila town and checked in to the Coral View Hotel and dive centre. There was frantic activity as preparations were going on for a night dive to monitor coral spawning. We were standing on coral gravel at the back of the wooden dive-centre building in the punishing heat of the afternoon. Dominic, a PhD student of mine studying the fish communities of the reefs, was in charge of the expedition. I was tired after the journey, but Dominic had told me that the corals might spawn that night, which was something I had never seen before. I was raring to go.

Ed, a tall, dark-haired Aussie in charge of dive safety, explained he would have to do a diving check at the start of the dive. I set about getting my equipment prepared, and, as it got dark, I struggled into my wetsuit and hauled on my cylinder and other equipment and plodded out along the jetty, mask and fins in hand, to the bar which had steps running down into the sea. Ed waited in the water as I stood on the steps holding the rail in one hand while pulling on my fins with the other. After some contortions everything was on, and I stepped into the sea in only a metre or so of water and then slid forward face down in the Caribbean Sea to swim after Ed.

It was disorientating at first. It had been a year since I had been to Utila, this was my second visit and the coral standing around me looked unfamiliar. It was also extremely warm, like a hot bath. I didn't remember it being this warm, almost uncomfortably so, before. I followed Ed to a gap in the limestone where we broke on to the reef face which sloped away below us.

One of the things about night diving is that your world is contracted down to what can be seen in the torch beam. It is somewhat like having tunnel vision, so when something hit me in the face I had no idea what it was. It happened several times in a row and finally I got my torch on a small silvery fish whirling in a vertical loop as though it was a toy on a string being spun by a child. I became conscious that the water was swarming with life. In the maelstrom I could only identify the odd creature by tracking it with my eyes. There were segmented worms called polychaetes; salps, tube-like gelatinous animals jetting through the water; and millions of copepods, small crustaceans, like dancing fleas, suspended before my eyes. I saw a *Phronima* and did a double take, this was a typical animal of the twilight zone, a large amphipod crustacean which always reminded me of the monster from *Alien*, living in a hollowed-out salp house. It tumbled and spun away into the darkness. Looking out into the blackness with the torch turned off, I saw spots of blue-green light blink on and off. Bioluminescence.

The reef was frantic with activity. Like us, the animals were waiting for the corals to spawn, producing a feast of bundles of eggs and sperm for every resident and visiting critter. Coral spawning only takes place once a year, in line with a specific phase of the moon, and always at night. Max, a student from Oxford, beckoned me over to a colony of *Orbicella*, also known as mountainous star coral. The colony was a deep brown-green colour and covered in polyps – the individual plump, anemone-like animals that make up the coral. As we watched over a period of 20 to 30 minutes, the coral polyps began to bulge. Gradually a white mass became visible through the mouth of each coral polyp and was held there. I was anxious that for some reason the corals would not release what would become their progeny. The mad swirling of the swarms of plankton buzzed before our eyes. All of a sudden, as if some invisible cue had been given, each of the coral polyps gave birth to a perfectly spherical white globe, hundreds being released by a single colony. They slowly drifted up in clouds. All around us the same was happening as the corals tried to swamp the predators with their precious eggs by sheer numbers. I had never seen this before and watched with fascination as the spheres rose like miniature hot-air balloons, disappearing towards the surface, borne by the current. This was an event that had been repeated for perhaps hundreds of millions of years. Beyond our sight the bundles of eggs and sperm would break down and the eggs would be fertilized to produce planula larvae. These resembled tiny slugs covered in fine, beating hairs which propelled them along. I had read about coral spawning and seen it on natural history programmes but had never seen it with my own eyes. I was entranced and felt very privileged to see the event that is so profoundly important for the continuation of coral reefs generation on generation. I also found myself willing the tiny propagules to escape into the dark and avoid the swarms of predators that beset the breeding corals.

We found our way back up the reef. The break through to the reef flat was marked by an enormous colony of *Dendrogyra*, a coral that looked like a series of furry tube-like towers. The change in temperature was very noticeable as we passed through the thermocline, the depth at which warm surface water lay on top of cooler water below. The change in temperature was so marked that the warm water sat in a distinct layer above the colder, denser water below. Mixing them blurred our vision as water of different densities swirled together. Again I found myself in what felt like a warm bath. As we swam through the shallow water, I began to notice many of the corals were a strange colour. Instead of a rich brown or green they were pale green, blue or even pink. It was when I started to see white colonies that I realized, with a deepening sense of unease, that what I was looking at were the early stages of coral bleaching.

To understand what mass coral bleaching is and the serious threat it poses to coral reefs you have to understand something about the biology of reef-forming corals. Stony corals, or *Scleractinia*, are ancient animals that have been building reefs in our oceans for over 200 million years. They evolved during the Triassic period between about 252 and 201 million years ago, when the first dinosaurs roamed the land. Their success is related to an intimate relationship with microscopic algae called zooxanthellae, which live in the coral tissues and give them their green or brown colours. Zooxanthellae photosynthesize, just like plants on land, and produce more than 90% of the nutrients that corals need to grow. The corals, in return, provide the zooxanthellae with shelter from grazing animals that would eat them, and supply them with nutrients, such as nitrogen, which they need to grow. This symbiotic relationship is central to the ability of corals to grow and form reefs. The corals extract calcium carbonate dissolved in seawater and secrete it as aragonite to form their skeletons.

In Honduras, these corals exist in a bewildering array of forms, from branching treelike skeletons, to large cylindrical growths (like *Dendrogyra*), to encrusting mats, delicate plate-like forms and bubbling mounds (like *Orbicella*), to massive brain-shaped robust structures (like *Diploria*). As with *Lophelia pertusa*, reef-forming corals are formed of colonies of individual polyps produced through budding, a form of asexual reproduction. As the corals grow and die, their skeletons are incorporated into the reef, forming the massive limestone structures that are so famous. The Great Barrier Reef of Australia is the largest structure formed by living organisms on Earth and can be seen from space. The second-largest barrier reef system is the Mesoamerican Reef in the Caribbean, and the reefs we were studying around Honduras form the southernmost parts of this system.

Mass coral bleaching occurs when water temperatures exceed their normal range over a period of weeks. The corals have to be exposed to sunlight, which gives a clue as to why bleaching happens. A natural by-product of photosynthesis is oxygen and the microscopic algae floating in the ocean as plankton produce close to half of the oxygen generated by the biosphere and released into the Earth's atmosphere. However, when temperatures get too high the microscopic algae, the zooxanthellae, go into photosynthetic overdrive producing what are called reactive oxygen species (ROS). These include superoxide and hydrogen peroxide, the bleaching agent used to dye hair. These chemicals are highly reactive and damage proteins, DNA and other important molecules in the coral tissue. In humans ROS have been implicated in ageing and in causing cancer because of their damaging effects on our own cells. The corals, in an effort to reduce this damage, eject the zooxanthellae and lose the brown or green colouration that the algae confer to their soft tissues. The bleached white colour typical of this condition results because the white aragonite skeletons of the corals are visible through their now-transparent tissue. If temperatures remain anomalously high for several weeks or more the corals essentially starve to death or succumb to disease. It is called *mass* coral bleaching because it occurs over vast areas. Corals bleach at local scales in response to disease or pollution and it may be a natural mechanism that has evolved to limit damage by an upset in the delicate physiological waltz between the host and its symbiotic algae.

While I was running around the seashores of Ireland as a child in the late 1970s, the first mass coral bleaching events were detected on the Great Barrier Reef in Australia. In 1997/1998 a single mass-bleaching event across the Pacific and Indian Oceans killed 16% of all the world's coral reefs in a single go. In places in the Indian Ocean the mortality was more than 90%. Imagine if 16% of the world's rainforests were lost in a year. There would be uproar. The 1997/1998 event almost went unnoticed in the developed world. As time has progressed, bleaching events have become more frequent and more widespread. The warning signs in Honduras in 2015 heralded the worst mass-bleaching event yet, which occurred across three consecutive years, continuing into 2017. This was a global event in the sense that it affected almost every region in the tropics where coral reefs occur. This was the third such event, the 1997/1998 one being the first, with a second in 2010, which was particularly severe in South East Asia but also affected corals in the Caribbean and Australia.

As I am writing we are still counting the costs of the latest event where 70% of the world's coral reefs experienced levels of heat stress associated with mass bleaching and death of corals. Along with many other people, I watched the depressing news reports on the devastation wreaked on the northern part of the Great Barrier Reef where entire fields of bleached white coral were shown. The BBC's *Blue Planet II* graphically showed time-lapse films of corals losing their colour and fading suddenly from a healthy green to skeletal white. Elsewhere, many island states in the central and south-west Pacific were devastated by bleaching. Christmas Island, part of the island nation of Kiribati, lost 80% of its corals. Further north, the Ryuku Islands of Japan were severely affected with the largest reef, Sekiseishoko Reef, suffering 90% bleaching leading to the death of 70% of the corals. China, Vietnam and Taiwan were also affected. In the Caribbean, as well as the bleaching we saw in Honduras, the same damage was also observed in Cuba, the Turks and Caicos Islands, the Bahamas, Haiti, the Dominican Republic and Mexico.

In 2016, a year after I witnessed coral bleaching first-hand with the Thinking Deep project, my own team were diving in the Chagos Archipelago, a remote set of coral atolls lying in the central Indian Ocean. Chagos has the reputation for being among the most pristine of coral reefs globally. This is probably because it has no human population except for a large American airbase on one of the atolls. In 2010 the British government declared the entire maritime zone as protected from all forms of commercial fishing, a move that was controversial at the time because of an active tuna fishery in the region and also because of a sovereignty dispute with Mauritius. This amounted to an area of 550,000 square kilometres and included not only the world's largest atoll, Great Chagos Bank, but also a significant proportion of the sea-mounts of the Indian Ocean. The team, which included Dominic and also a postdoctoral scientist, Catherine Head, were diving on the reefs across the archipelago on a project funded by the Bertarelli Foundation to examine reef biodiversity. I was working in Oxford and began to receive the bleak news via email that Chagos had been severely affected by the mass-bleaching event. Live coral cover had fallen from 30% to 12% at water depths of down to 15 metres and in places from more than 40% to 7%. The soft corals had been all but obliterated from shallow water. What had been a vibrant, colourful reef was now grey eroded limestone covered by an algal fuzz. The reef had gone from a healthy state, where it was growing vigorously, to a situation where it was eroding. When Cath and Dom returned they were visibly depressed by the whole experience. Cath had worked on the coral reefs of Chagos throughout her PhD. To see such a thriving and beautiful reef reduced in many areas to barren deserts in a matter of a few years was crushing.

Mass bleaching of coral reefs is a result of global warming driving up the temperature of the ocean. Carbon dioxide emissions cause the atmosphere to trap the heat of the sun and the ocean has absorbed over 90% of the heat resulting from this process. About a quarter of the $CO_2$ emissions come from burning fossil fuels for electricity and heat production, another quarter from agriculture and deforestation, and about a fifth from industrial processes. Transport by road, rail, air and by shipping contributes another 14% with the remainder made up from heating and cooking in buildings and through actually generating fossil fuels (extraction, refining, etc.). Global warming induced by human activities is a well-understood phenomenon. As early as the 1890s, the Swedish scientist Svante Arrhenius had realized that increases in atmospheric $CO_2$ would increase global temperatures. A colleague of his, Arvid Högbom, calculated that industry was already producing comparable amounts of $CO_2$ to natural sources. In the late 1950s, Charles Keeling, a US scientist working at the Scripps Institute of Oceanography, set up a $CO_2$ observatory on the slopes of the world's largest volcano, Mauna Loa in Hawaii. The location was important because it was in the central Pacific, away from continental sources of pollution and was sufficiently high to avoid measuring local $CO_2$ emissions. Within a few years he proved that $CO_2$ levels in the atmosphere were increasing. Anybody denying that human $CO_2$ emissions are driving up the temperature of the climate is ignoring what now amounts to a very large body of peer-reviewed scientific evidence, which only strengthens as time goes on. Yet still, politicians deny the overwhelming evidence of human-driven climate change. An example has been the widely reported removal of terms such as 'climate change', 'greenhouse gases' or 'global warming' from the Environmental Protection Agency website in the United States following the election of Donald Trump as president. Even the IPCC reports that summarize the results of scientific studies on climate change have been criticized for having important data and text removed during negotiations between governments on the final wording of these important documents for policymakers.

That world leaders and their governments are still in denial about climate change, its scale and the devastation it is already wreaking on the natural world, is a crime against nature, humanity and particularly future generations. While some politicians and their state apparatuses are suppressing the findings of scientists, we are facing nothing short of a global emergency. In 2007 I attended a workshop in the Philippines. In a whitewashed meeting room in the humid heat of the tropics, scientists discussed population declines of corals as a team from IUCN drew maps and filled out reports on the current status of each species. At the end of the meeting we all gathered to assess what had been found. Of the 700 species of reef-forming corals for which there was enough information to make an assessment, *a third were threatened with extinction*, mainly as a result of the impacts of mass bleaching. The audience of scientists sitting there in a garish array of Hawaiian shirts, T-shirts and shorts were visibly in shock. It was incredible that in just a few decades, ocean warming had pushed so many species towards extinction. Coral reefs had become the most threatened ecosystem on Earth.

The experts in the room decided to send a paper based on the results of the workshop to the journal *Science* to try and wake up the world to the plight of coral reefs. When I returned to the UK I got together with Mirella and Sophie from Communications INC and we planned a scientists' meeting to try to publicize the plight of corals prior to the Copenhagen climate change conference which was to take place in 2009. The Zoological Society of London was also focusing on the plight of coral reefs and produced a short film, *Corals on the Edge*. It was based on a young girl from the future talking about a piece of coral skeleton she found on the beach and her grandfather telling her that when he was a child there was a great reef built by the corals full of colourful fish and other life. Every time I heard it I found myself welling up as I couldn't help think about my own children and what was being lost to them in the future.

The scientists' meeting was held in the Royal Society, London, in July 2009 with experts on coral reefs and climate change from all around the world. The meeting was co-chaired by Sir David Attenborough. If anything, my sense of alarm grew through the meeting, particularly with respect to the consequences of ocean acidification. As well as absorbing 90% of the excess heat accumulating as a result of climate change the oceans are absorbing about a third of the $CO_2$ we produce. Both of these actions contribute to less warming of the atmosphere. However, the absorbed $CO_2$ is converted to carbonic acid, and this in turn alters the equilibrium between calcium carbonate and calcium bicarbonate in seawater. Calcium carbonate, in the form of the mineral aragonite, is what corals build their skeletons out of. As seawater was becoming less alkaline the amount of carbonate in it was dropping. Ken Caldeira, a New Yorker and a vociferous climate change scientist from the Carnegie Institution for Science in California showed a series of colourful maps where the ocean changed over time from purple to blue to yellow and then red as it moved from a carbonate-saturated state to one where it was corrosive to calcium carbonate. The picture Ken painted was grim. As he stepped forward in time over the next hundred years or so, the area of ocean where carbonate levels were suitable for coral growth shrank dramatically.

The level of acidity is measured on the pH scale, a measure of how acidic or how alkaline a solution is. The H stands for hydrogen ions, which are responsible for the level of acidity of a solution. The scale is a negative or reciprocal one, so concentrated hydrochloric acid has a low pH value of 0.1, vinegar is about pH 3, tap water is neutral at around 7, seawater prior to the industrial revolution was slightly alkaline at about pH 8.2, household bleach, which is very alkaline, is about pH 12. Seawater has changed in pH by about 0.1 units which sounds a vanishingly small amount, but this is a logarithmic scale, used to represent a large range of quantities. The values of pH represent changes of hydrogen ion concentration in orders of magnitude rather than on a linear scale so that the change of 0.1 pH units actually means a 30% increase in acidity. Such a large change in hydrogen ion concentration profoundly affects the way chemicals behave in seawater and in turn impacts the organisms living within it. Studies were already suggesting that the growth of reef-forming corals in areas such as the Great Barrier Reef was slowing down. That our burning of fossil fuels and other activities have actually changed the chemistry of the ocean is almost inconceivable. The volume of the ocean is about 1.3 billion cubic kilometres, an almost unimaginable mass of water. However, humankind has achieved this dubious distinction, making our impact now one that could be recognized over geological timescales. Welcome to the Anthropocene, the age in which human activities have become a significant influence on our climate and environment.

The meeting was alarming and depressing. Charlie Veron, a veteran coral scientist, a man thought to be responsible for naming 20% of the world's reef-forming corals, did not hold back. If something was not done soon, coral reefs would be gone. The scientists all signed up to a Statement of Concern and a group of us decided to put together a paper: 'The coral reef crisis: the critical importance of <350ppm $CO_2$'. Coral reefs were the canary in the coal mine for climate change. Sir David Attenborough led the press conference following the meeting and summed up the problem eloquently: 'If we do nothing coral reefs will end up as slime-covered rubble piles.'

It was a straightforward statement of fact. The press hung on to his every word and it was a spellbinding performance – which was just what we needed to get the world to sit up and pay attention.

The Copenhagen climate change conference, which took place a few months later in December 2009, was a huge jamboree.

Top: Dense aggregations of the yeti crab (*Kiwa tyleri*), also known as the 'Hoff crab', on the southern site at the East Scotia Ridge, Southern Ocean, 2,397m depth. NERC CHESSO project.

Bottom: Unidentified vent octopus, possibly a *Vulcanoctopus* species, southern site of the East Scotia Ridge, Southern Ocean, 2,394m depth. NERC CHESSO project.

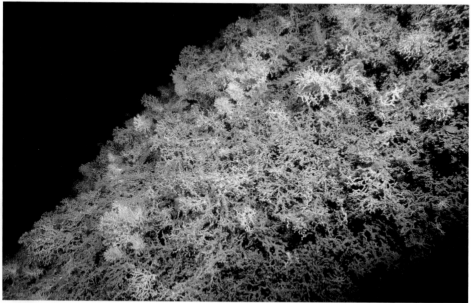

Top: Actinostolid sea anemones and stalked barnacles (*Vulcanolepas scotiaensis*), southern site of the East Scotia Ridge, Southern Ocean, 2,396m depth. NERC CHESSO project.

Bottom: *Lophelia pertusa* forming a cold-water coral reef on the Anton Dohrn seamount, north-east Atlantic, ~400m depth. NERC Deeplinks Project.

Top: Live colony of *Solenosmilia variabilis,* part of cold-water coral reef on Coral Seamount, South West Indian Ridge, Indian Ocean, ~1,000m depth. The snail is a calliostomid (top shell) probably eating the coral polyps.
IUCN/NERC Seamounts Project.

Bottom: Orange roughy (*Hoplostethus atlanticus*), Melville Bank Seamount, South West Indian Ridge, Indian Ocean, ~1,000m depth. IUCN/NERC Seamounts Project.

Top: Spiny lobster (*Jasus* cf *paulensis*), Sapmer Bank Seamount, South West Indian Ridge, Indian Ocean, ~200-300m depth. IUCN/NERC Seamounts Project.

Bottom: The flattened, transparent larva of a spiny lobster showing a remarkable lack of resemblance to the adult. Captured in a pelagic trawl above the South West Indian Ridge, Indian Ocean. IUCN/NERC Seamounts Project.

Top: Beautiful bubblegum coral (*Paragorgia species*) photographed on Atlantis Bank Seamount, South West Indian Ridge, Indian Ocean, ~700-800m depth. IUCN/NERC Seamounts Project.

Bottom: The spiny regular urchin *Dermechinus horridus* on coral on the summit of Atlantis Bank Seamount, South West Indian Ridge, Indian Ocean, ~700m depth. The lobster on the right is a *Projasus* species. IUCN/NERC Seamounts project.

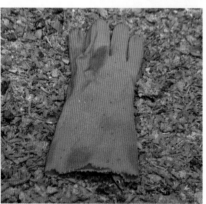

Top: Rock outcrop on Atlantis Bank Seamount, home to sea fans, sea anemones, sponges, the urchin *Dermechinus horridus* and the glass sponge *Euplectella*. Inside each sponge was a male and female shrimp, a new species. South West Indian Ridge, Indian Ocean, ~700m depth. IUCN/NERC Seamounts Project.

Bottom left: Fisherman's glove left behind on Sapmer Bank Seamount, South West Indian Ridge, Indian Ocean. Note an eel has taken up residence under this particular piece of marine litter. IUCN/NERC Seamounts Project.

Bottom right: Dreadful scene of floating plastic, polystyrene and other waste of human origin stretching out to the horizon. Island of Utila, Bay Islands, Honduras, 2015. Photo © Erika Gress.

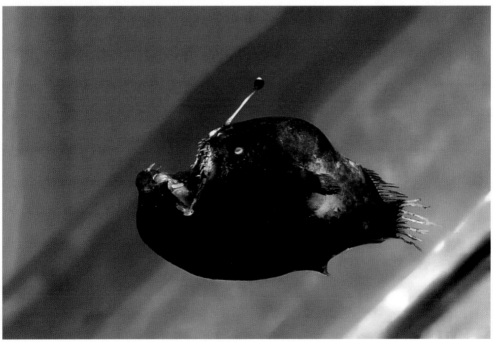

Top: Antarctic fur seals (*Arctocephalus gazella*), Island of South Georgia, South Atlantic. Photo © Alex Rogers.

Bottom: The black devil anglerfish, *Melanocetus johnsoni*, sampled from the twilight zone above the seamounts of the South West Indian Ridge, Indian Ocean. IUCN/NERC Seamounts Project.

Top: The author as a larval marine biologist on the rocky shore at Cloonagh, Sligo west coast of Ireland, sometime in the 1970s. In the background are my father and grandfather, and my uncles, John and Peter, launching my grandfather's fishing boat. Photo © Alex Rogers.

Left: Ex-PhD student Chong Chen takes great pleasure in subjecting his supervisor to the traditional 'Neptune's soaking', after a first dive in the *Shinkai 6500* submersible. Photo © Nicolai Roterman.

The Zoological Society of London sent a delegation along with many NGOs. Everybody was expecting a momentous decision because the evidence that climate change was devastating the planet was now incontrovertible. Instead we faced crushing disappointment when a non-binding and very vague statement was produced by the USA, China, India, Brazil and South Africa. The frustration resulting from the outcome of the meeting was palpable within the scientific community. For many of my NGO colleagues the sense of loss was like a bereavement. For others it was seen as a betrayal of the citizens of Earth by uncaring and remote governments only interested in economic gain at the expense of all else.

The politicians in Copenhagen had procrastinated while the house continued to burn down.

Coral reefs in the Caribbean faced other problems too. In Honduras we slipped into the routine of walking down through Utila town to the Coral View dive centre early in the morning. This involved navigating packs of semi-feral dogs, piles of rubbish, speeding tuk-tuks and mopeds and a variety of other colourful hazards. Sometimes I'd get picked up by Marie, the chief diving instructor, on a quad bike. We began by taking further dives off the bar at Coral View. Again I was struck by just how warm the water was as we swam through the break in the coral wall under water just a few metres deep towards the steps. As I looked up I spotted a large colony of elkhorn coral, named because the irregular broad branches resemble the antlers of an elk or moose. It was noticeable because it was such a rare sight on the reefs around Utila. However, although about two thirds of the colony was green, stretching up from the main axis of the coral and across several of its branches were bright-white patches. At first I thought this was more coral bleaching, but I then realized I was probably looking at white band disease.

Elkhorn coral and stag's horn coral once dominated the shallow waters of the Caribbean. Stag's horn coral forms colonies of irregular pointed branches (which, you won't be surprised to hear, resemble the antlers of a stag). Prior to the late 1970s these corals were the prominent reef-constructing corals in the shallow waters of the Caribbean. A quick browse through a book by Hans and Lotte Hass, two diving stars of underwater films in the 1950s and 1960s, or a viewing of one of Jacques Cousteau's films from the Caribbean will reveal just how common they were. There are photos of Lotte hanging on to the broad-bladed branches of enormous elkhorn coral colonies (which is not to be advised with any coral as it kills the delicate polyps). On *Mysteries of the Hidden Reefs*, aired in the mid-1970s, Cousteau's divers are seen among submarine forests dominated by stag's horn coral searching for specimens for the Discovery Bay Marine Laboratory in Jamaica. The reefs are now gone, collapsed into algae-covered piles of rubble and along with them most of the marine life that teemed within the bay. Elkhorn coral could form huge beds in shallow water as complex in structure as tropical rainforests. Sheltering among the branches and understorey of the coral was a tremendous variety of reef fish and other creatures. The elkhorn corals almost seemed to be reaching out for the sun above the surface of the sea.

The first reports of large-scale coral mortalities associated with disease were from Florida, where a disease called white pox killed off large numbers of colonies of a coral called *Mycetophyllia ferox*, a species which forms semi-encrusting ridged plates on the seabed. It was in the early 1980s that white band disease first appeared, wiping out large areas of both elkhorn and stag's horn corals. The fields of white corals would quickly turn a dirty green as they were covered in algae and then the framework would collapse. There soon followed other diseases, including black band disease, white plague, red band disease and yellow-blotch. Epidemics were not only hitting the corals. In 1983, a disease that affected the long-spined sea urchin, *Diadema antillarum*, appeared off Panama. *Diadema* was a key species on Caribbean coral reefs as they grazed algae which could potentially compete for space with the coral and smother it. This importance increased after populations of algae-grazing fish, such as parrotfish, were removed by overfishing. Within 12 months of the disease being detected it had spread to urchin populations across the whole Caribbean. 95%–99% of the urchins died, with the result that the algae began to dominate space on reefs. In Discovery Bay, Jamaica, algal cover leaped from 30% to over 70%. Corals were simply overgrown by algae, or if they died, algae took their place. The replacement algal communities had a lower diversity of fish and other animals associated with them, they were nowhere near as rich as the coral communities they had ousted.

The situation was rendered even more catastrophic by the 2005 hurricane season, which had the worst storms ever recorded in a single season at the time. Hurricanes such as Katrina, still one of the deadliest and most-costly hurricanes recorded to date, flattened large areas of already-stressed coral reef. The result of this death by a thousand cuts was a change in the structure of Caribbean reefs. Scientists collected together studies where the three-dimensional complexity of reefs was measured. At its simplest this is done by laying a length of chain across the reef. The distance between each end of the chain divided by the length of the chain laid on the reef and going up and down in all the nooks and crannies, gives a complexity score. A score of 1 would indicate a flat surface and then increasing scores would indicate a more complex surface. It was found that from the late 1960s to 1985 the reefs across the Caribbean had undergone a marked reduction in complexity. It then stabilized until the massive bleaching event in 1998 when the collapse recommenced and complexity began to decrease still further. As with reefs formed in deep water, three-dimensional complexity of a habitat is important to many of the other animals associated with the reef. The overall collapse in structure of the reefs meant less habitat for many other species.

What were the causes of the epidemics that swept Caribbean coral reefs? That's not entirely clear in many cases. White band disease appears to be associated with several different bacteria, but no one organism has been identified as the definitive pathogen. What is known is that the occurrence of disease appears to be associated with high sea surface temperatures. It has been discovered that corals have a complex community of microorganisms associated with them. These microbes live in the mucus coating the surface of the corals, in the coral tissues, in the coral equivalent of the gut and even in the coral skeleton. This microbiome includes a wide variety of bacteria, viruses, fungi and protozoans. This menagerie of microbial life appears to be intimately associated with many functions in coral colonies, including nutrition of the coral hosts, nutrient recycling and disease prevention. There is increasing evidence that when temperatures reach stressful levels there are changes in the coral microbiome, including an increase in disease-causing bacteria.

For other diseases, such as white pox, a more direct cause has been identified. In this case the bacteria *Serratia marescens* is the pathogen. This bacterium is associated with the human gut and therefore with sewage. The likely route of the disease is the large volumes of untreated sewage released into the ocean which brings *Serratia* into contact with the corals and has other detrimental effects, such as encouraging the growth of algae. The *Diadema* epidemic originated near to the mouth of the Panama Canal so there is a suspicion that perhaps the disease was non-native and introduced from the discharge of ballast water from ships traversing the canal.

A few days after seeing the majestic but diseased elkhorn coral colony, I dived on the coast adjacent to the airport in Utila. We dropped onto the shallow reef and then swam to the drop-off and down to 20 metres depth to place sediment traps, weighed-down pieces of plastic piping. Our first encounter was a large turtle lying on the reef busily munching sponges off the rocks. It was completely unconcerned by our presence. We placed our traps and turned, almost bumping into a large spotted eagle ray obviously curious and passing close by to eyeball us. I love these animals, with strange boxy heads, eyes on either side and pointed

171

wings a beautiful dark-blue colour scattered with bright-white spots and followed by a long, pointed tail. After watching the ray gracefully swim into the blue, like a swan in slow motion, we swam up the reef to finish placing our traps. We passed over stony corals and swaying sea fans, all greens and purples with the odd splash of red, yellow or orange from sponges. Blue-black triggerfish swam past overhead and colourful parrotfish nipped in and out of crevices foraging for algae. After dropping the last traps, we had to find a tape left behind by one of the dive teams the previous day at five metres depth. This brought us up onto the reef crest, a less lively area with outcroppings of coral, limestone and pools of sand.

As we swam along searching for the tape a remora appeared and began to follow Max, one of the students working with us. These are quite sizeable fish with a large sucking disc on the top of the head and a black stripe running down the body. They have pointed triangular fins and somewhat resemble a shark which is what they usually attach to. This one decided that no matter what it was going to attach itself to Max. Despite his shushing it away several times as it swam around his head, it eventually settled to hitch a ride on his air tank. We came upon the tape, which was bright yellow and so hard to miss. As I looked up my breath was almost taken away. There in the shallows was the ghost of a reef. The water was exceptionally clear and in the greeny-blue light reflected from the reef and sand were the stony dead skeletons of numerous elkhorn coral colonies. They were scattered randomly as far as I could see, metres apart, and only eroded stumps, like an accusation of a recent murder. I was devastated because here was the proof of what had come to pass in the Caribbean over the period of just 30 to 40 years, less than a lifetime, the blink of an eye. This area had once obviously been dominated by the great branching corals, a shallow forest, probably teeming with fish and the under-canopy richly carpeted in other life. Now what was left looked like the aftermath of an H-bomb underwater. The odd structure still standing, dead, decaying and grey. It was a graphic example of the flattening of the reef. I still hold the picture of that field of dead coral in my mind and I see it whenever I feel as though the battle to save our planet is too tough. It keeps me fighting.

We swam back, feeling subdued, with the remora trying to cheer us up. It had decided that now I was a more comfortable ride than Max and after looking at me through my mask several times it disappeared. Max told me when I surfaced that it had hitched a ride on my tank most of the way back to the boat.

As we moved into the expedition my dive team began to prepare for the really deep dives on the reef. The rebreathers were checked and checked again, the Sofnolime canisters that scrub exhaled carbon dioxide from the diver's breath were carefully packed and cylinders of 'bailout' gas, used in emergencies and gas for planned decompression prepared. At an early stage, Dan, a collaborating scientist, and I had discussed working on the deeper parts of the reef, the mesophotic zone. Dan was a wiry energetic character, and worked in a non-governmental organization providing educational experiences for students, Operation Wallacea, which had been working in the Caribbean for several years. He'd identified Utila in Honduras as a place we might be able to develop technical diving capabilities to dive to the depths required, up to 90 metres. This was because of the presence of a technical dive school at Coral View and the ability to ship oxygen and helium to the islands, which would be necessary to develop such a programme.

Mesophotic reefs are the deeper parts of tropical coral reefs which lie in shallow waters. They are found between the greatest depths of conventional scuba diving – 30 metres – down to the greatest depths where zooxanthellate or light-harvesting corals can live. In some parts of the world this can be greater than 150 metres. Mesophotic reefs fall into a crack in the pavement in terms of access technology. At depths shallower than 30 metres scuba divers can carry out research by undertaking surveys, collecting samples and setting up experiments. Deep-submergence equipment, of the type I have typically used for exploring hydrothermal vents and seamounts, is usually used at depths from 300 metres to thousands of metres deep. It is difficult to use in shallower water because it needs to be deployed from large ships, which cannot move too close to a reef for obvious reasons. As a result, mesophotic reefs had become neglected by scientists. However, in the context of climate change, impacts like mass coral bleaching and other, more direct, impacts on reefs, the mesophotic zone might be important in terms of reef resilience. This is because deeper waters are sheltered from some of the more extreme conditions that affect shallow-water reefs. They are less exposed to bright sunlight, as seawater absorbs light. They are also at lower temperatures and are exposed to less wave action, especially during storms. Considering this, scientists realized that if these deeper reef environments shared species of corals, fish and other animals with shallower reefs they might act as sources to recolonize shallow waters following an extreme event like a mass bleaching. You can imagine the mesophotic reefs as a kind of natural seed bank for shallow-water reefs, a refuge from the stresses being exerted on shallow waters by us humans. This became known as the Deep Reef Refugia Hypothesis (DRRH). If this hypothesis was true, then coral and fish distributions should stretch across both shallow and deep waters. There should also be evidence that reef stressors declined with depth. The Thinking Deep expedition was all about looking at the possibility that the mesophotic reefs around Utila might replenish shallow-water reefs in the event of a disaster like mass coral bleaching.

Over the weeks we began to build up a picture of the meso-photic reefs around Utila. Two discoveries made during the expedition were very encouraging in terms of supporting the idea of deep reefs as refuges. First, the coral communities themselves appeared to be comprised of a group of shallow-water specialist corals found at depths of five metres or so on the reef. Many of the other corals were found across a range of depths on the reef from five metres downward into the mesophotic zone. Only a few species were restricted in their distribution to depths greater than 25 metres. This meant that if shallow waters of the reef were heavily disrupted by, for example, a mass-bleaching event, the corals that showed little preference for any particular depth might replenish the shallower areas of the reef. In the very shallowest waters this may mean the loss of the shallow-specialist corals, but at least there might still be coral reef. Second, we saw some success with transplanted corals. Jack, a student on the expedition and one of our deep divers, had collected some fragments of living corals from shallow and deep water, mixed them up and then transplanted them across the depth gradient. This meant a mixture of corals of shallow and deep origin at all depths across the transplant site. This experiment coincided with bleaching on the reef and demonstrated that the corals transplanted at greater depths showed lower mortality from the bleaching event than in shallow water. This fitted with the idea that corals at greater depths were sheltered to some extent from some of the stressors affecting shallower reefs. This was a necessary prerequisite for the Deep Reef Refugia Hypothesis to work and explained why in places some coral species had been completely lost from shallow water but remained at mesophotic depths. In 2016 my team had found the shallow-water reefs in the Chagos Archipelago devastated by mass bleaching that had resulted from anomalously warm waters over a period of two years. However, they had also deployed a small remotely operated vehicle and discovered that in some locations there was dense coral cover between 30–60 metres depth. This was immensely encouraging.

However, during the expedition our work showed that not all human-induced stresses declined with depth on the reef. The work on fish communities in Honduras aimed to show how the diversity, abundance and biomass of species changed with depth. During the deep dives our team discovered large aggregations of lionfish, a predator native to the Pacific and Indian Oceans which has invaded the Caribbean. They are spectacular in appearance, striped in white and red bands with an array of long fins which look like vertical prayer flags arranged like fans. These fins are armed with venom to fend off would-be predators. The head is boxy with a large mouth surrounded by feathery frills and a rounded diamond-shaped antenna above each eye.

In the early 1980s, either deliberately or accidentally, lion-fish were introduced into the northern Caribbean. They spread rapidly through the reefs of the region feeding on native fish species. I was aware of lionfish from diving on the shallow reefs of Honduras. We would occasionally see the fish hovering just above the coral but as soon as we approached they would invari-ably disappear into a crevice. This was because in Honduras, as elsewhere in the Caribbean, lionfish were culled in shallow water by scuba divers and had learned to avoid them. At mesophotic depths, however, our divers discovered that the lionfish were much more numerous. Instead of the odd timid individual found hiding in crevices in shallow waters, the deep divers were coming back with stories of a dozen or more lionfish gathered in aggre-gations off the reef walls. Alison, the expedition photographer, came back with photographs of six or more lionfish in a single location. As with other reef fish, young lionfish begin their lives in shallow water among seagrasses in the reef lagoon or in mangrove swamps before moving out on to the reef proper. Then, as they grow larger, they move into progressively deeper water. Because they are largely protected from scuba divers with spears below 30 metres depth, the impact of this invasive species was likely to be greater than in shallow water. This was one threat that seemed to increase with depth rather than decline.

A year later I was diving off Bermuda at a site called

# The Deep

North-North-East on an expedition for the Nekton Foundation. I was in one of two submersibles, a research submarine built by Triton, a company based in Florida. In the other sub was an adventurer-writer for *Forbes* magazine, Jim Forbes. He had driven cars at over 200 miles per hour and climbed the Matterhorn and that day he was heading for 1,000 feet in a submersible. He was nervously excited prior to the dive and we chatted about what to expect. I too was looking forward to this dive, as for me deeper water was *always* better. We dropped down the limestone slope of the Bermuda platform in the submersible *Nemo* with Jim following in *Nomad*. The slope steepened somewhat as we approached 300 metres depth and we found ourselves travelling over terrain that was formed by limestone ridges and platforms covered with eroded rhodoliths, irregular stony concretions formed by algae. The rough bottom was covered in a rich carpet of twisted wire corals resembling a great mattress of bedsprings, two metres or more in height. Here and there were yellow sea fans and colonies of a very delicate white coral that might have been hydrocorals, also known as stylasterids. Clouds of pink fish with yellow tails scattered before us, with the larger adults sporting a bright yellow tail and dorsal fin. With the help of scientists from the Bermudian government we had tentatively identified these as rough-tongued bass. They were by far the most common fish below 100 metres depth. I then spotted a bright golden-red, barrel-shaped fish with a bony head and star-like pattern around the eye. I got very excited as I first thought it was an orange roughy, but it couldn't be because we were too shallow. The fish hovered in front of a small hole and turned around to face us presenting a quite narrow profile. I realized it must be a Darwin's slimehead, a close relative of the roughy but not recorded before in Bermuda.

As we moved down to 300 metres to get the magic 1,000 feet for Jim and to begin surveys, I spotted, unbelievably, a lionfish, hovering just above the bottom. I was sure these animals had never before been seen at such depths. We were in the true deep sea, below 200 metres deep, in the sub-mesophotic zone. It turned out this was a record depth for lionfish in the Caribbean and North Atlantic and a year later we published these observations in the first scientific paper from the expedition. As we continued to dive over the next few weeks we found more and more of the lionfish at depths of 200 metres or more. I began to realize how these animals were such effective predators. At a distance the fanned fin rays and striped pattern of the body broke up the outline of the lionfish so it was very difficult to spot. This was particularly the case when it was among other fish. The lionfish often sat right in the middle of the shoals of rough-tongued bass which were completely oblivious to the killer sitting right there among them. Sometimes there would be two or three of the stealthy lionfish within a shoal of bass, no doubt lazily seizing the odd fish when they fancied. It was a very sinister feature of this beautiful but deadly invader of the Caribbean reefs.

So, coral reefs have become one of the most, if not *the* most, threatened ecosystems on Earth, more threatened than rainforests or other terrestrial ecosystems. This is as a result of climate change, which is driving mass coral bleaching, ocean acidification and an intensification of hurricanes and cyclones. Epidemics of disease also seem to be promoted by ocean warming, perhaps interfering with the balance of the delicate microbiome of the corals. Other threats to the intricate ecosystem of coral reefs include the urchin epidemic, which almost wiped out these spiny grazers of algae, as well as invasive species such as lionfish. Reading these pages, you may wonder whether there is any hope for coral reefs. The importance of these ecosystems cannot be overstated. They are home to a quarter of all marine fish species and perhaps nearly a million other species of organisms, the marine equivalent of a tropical rainforest in biodiversity terms. For many coastal communities of humans, they provide a barrier to extreme weather events, and are a rich source of food and in some cases, of tourist income. Yet these naturally constructed jewels only cover about 0.1% of the area of the ocean with most being located in the South Pacific. They are truly on a knife edge, with the effects of climate change in particular becoming progressively more devastating every year. There is truly the chance that by the time my children are adults and perhaps have had their own children, a matter of a few decades, coral reefs will be gone or will be in a tail spin to oblivion. They may become what a colleague of mine has called zombie reefs, still there but no longer growing and no longer producing offspring.

There are signs of hope, though. The discovery of the reservoir of biodiversity in mesophotic depths is one aspect of reef ecology that might provide reefs with a stronger resilience to climate change impacts in the future. Some reefs are naturally protected from stresses because of their location. They might be sheltered by cliffs from direct sunlight throughout the entire day or located near an upwelling of cooler water, preventing temperatures in the waters around them from exceeding the coral's thermotolerance. We are also beginning to understand that some coral populations appear to be more tolerant of stressful conditions than others. These often live in conditions that are naturally more extreme than other reefs, where temperatures are higher for example, and where the corals and their symbiotic algae have adapted to tolerate environmental stress. Establishing protected areas where there is a threat of human activities directly damaging coral reefs, such as fishing, is also playing a role in conserving these ecosystems.

It has been shown that where reefs are healthy, with large populations of fish, they are more resilient to climate change impacts. This has sometimes been because the presence of large populations of herbivorous fish prevents algae from outcompeting coral when a reef is recovering from a shock like mass bleaching induced by high temperatures. Scientists are also looking at alternative approaches to improving reef resistance to a warming ocean. This involves artificially breeding corals and perhaps, in the future, genetically modifying corals and the symbiotic algae that live in their tissues and provide them with food to tolerate higher temperatures or seawater with less calcium carbonate in it. Reef restoration is also being attempted around the world, although with varying degrees of success in maintaining aspects of reef biodiversity, such as fish stocks.

It has been estimated that the financial benefits coral reefs confer to humankind add up to nearly $10 trillion US dollars per year. This is the overall value conferred from food provision, coastal protection, tourism and the many other services coral reefs provide us. However, I believe that ultimately we have a moral obligation to prevent them from sliding into extinction. Diving on a coral reef is a breathtaking, out-of-this-world experience. The exuberance of life, the colours, the intimate relationships that have been created by over 200 million years of evolution are marvels. For many people such an experience is life changing and inspiring. We should do everything in our power to conserve these ecosystems for present and future generations. Its loss would be criminal negligence on a planetary scale. It is time that countries put aside their differences and investors and industry realize that with globalization and profitmaking comes responsibility. Unless serious efforts are made to decarbonize the global economy, coral reefs and many other treasured ecosystems on Earth will be lost forever. Action at all levels of society, from governments to what we do ourselves as individual citizens really matter to the survival of coral reefs in the future. This is a subject to which I will return in the final chapter of this book. You will be relieved to know that individual actions taken collectively really can make a difference to the future of our ocean.

# 6

# Plastics and Other Pollutants:

## *Out of Sight, Out of Mind*

The detritus of humanity is now everywhere in the ocean. Plastics, polystyrene, fibreglass – whatever packaging we use, a portion of it ends up in the ocean. Reports have been appearing in the press for years about a giant garbage patch in the centre of the Pacific Ocean, rivers of plastic flowing into the ocean in the Far East and piles of plastic and other waste as tall as a person on the beaches of cities like Mumbai. As disturbing as these reports have been they seem to have had little effect on our awareness of the issue. In the UK it wasn't until the BBC's *Blue Planet II* aired, a programme that graphically showed its audiences the results of our throwaway culture on marine life, that people seemed to sit up and pay proper attention. Now ocean plastic is a major theme at every conference I go to on marine conservation. However, it is not the only pollution problem faced by the ocean and some may even have serious implications for human health.

I first became aware of the plastics problem as a result of the work in a project I led in the southern Indian Ocean in 2011. The seamounts of the Southwest Indian Ridge lie between Madagascar/South Africa and Antarctica. They are remote and it takes a week to reach them on a modern research vessel travelling at 10 knots, the equivalent of about 11.5 miles per hour on land. As I have already related, we found a lot of debris on the seamounts of human origin, most of it lost fishing gear, although there was also evidence of garbage chucked over the side from these vessels: beer bottles, gloves, pieces of engine and bits of plastic plumbing for example.

During the cruise we sampled sediments using cores, a simple plastic tube, inserted into the sand or mud and slowly withdrawn, taking with it a cylindrical sample of the seabed. Doing this on seamounts, however, was a painful affair, as many of the sediments comprised large, unconsolidated grains, with bits of coral scattered throughout. As a result many of the cores emptied out as soon as the core tube was withdrawn from the seabed leading to hours of frustration.

Several months after this I was standing in the Ocean Research and Conservation laboratory in the Department of Zoology in Oxford when the phone rang. It was Lucy Woodall, a postdoctoral researcher on the seamounts project based in the Natural History Museum in London.

'I just want to run something past you and see what you think,' she said.

'Fire away.'

'I'm finding what look like plastic fibres in the core samples.'

'What do they look like?'

'Well, they are tiny fibres . . . They look like they are different colours, which might indicate different types of plastic. And they are in every sample.'

That got my attention immediately. No one had seen such quantities of plastic microfibres in the deep sea before. We had retrieved sediment samples down to 1,500 metres. If they were down there on the Southwest Indian Ridge, one of the remotest parts of the ocean, they would be everywhere in the deep sea. Then the penny dropped and horror gripped. This meant that plastics had reached *every* part of the ocean. In this moment I experienced what I can only describe as vertigo. The consequences of this discovery would be immense.

'Lucy, we have to be absolutely sure about this.'

'I've checked with Michelle. The fibres seem to be on the coral samples as well.'

This was even worse news. If the fibres were on the corals that Michelle Taylor, a postdoctoral researcher in Oxford, was looking at, it was quite possible that they were already affecting them in some way. If the corals had to clear the plastics from their polyps, the small anemone-like soft parts of the coral used for feeding, it might be significant in energy terms, slowing down the growth of the animals.

I put the phone down with a feeling of deep disturbance. What was the source of these plastic fibres and how were they getting down to the deep sea? Were they neutrally buoyant or were they negatively or even positively buoyant? The obvious answer was that the fibres were being spread out on the ocean surface by waves, tides, currents and the wind, and then being tangled in sinking phytoplankton and other material. In other words, it must be being trapped on marine snow, aggregates of organic detritus sinking from the sunlit surface waters into the deep ocean where it was the main source of food for animals in the water column and on the seabed.

There are several sources of microplastic particles and fibres. In some cases, the plastics are manufactured to be small, such as the plastic microbeads added to some personal care products, such as body scrubs and toothpastes. These are now being banned in countries around the world, and there are many biodegradable alternatives to their use. In other cases, they come from the breakdown of plastic materials on land or in the ocean. As large pieces of plastic are weathered by ultraviolet radiation in sunlight, oxidation and the mechanical impact of ocean waves, they break down into smaller and smaller particles, ending up as microplastics. The majority of microplastics in the ocean are thought to originate from the breakdown of larger pieces of plastic, but the relative quantities of plastic from different sources and where it goes is not well understood.

Another source of microfibres is from textiles. Approximately 9 million tonnes of textiles are manufactured a year of which about 30% are cotton, other natural fibres such as silk and wool make up 10% and the remaining 60% are synthetic. Most of these natural and non-natural fibres are treated with a cocktail of dyes and chemicals and can also accumulate pollutants when in seawater. Microplastic particles and fibres enter the ocean through storm drains, runoff from the land, through the outfalls of waste-water treatment plants and as dust blown through the atmosphere. The big stuff, so-called macroplastics (pieces 2.5–50 centimetres in length) and megaplastics (pieces more than 50 centimetres) end up in the ocean because many countries simply do not have the infrastructure to handle their solid waste. Something like 5 to 15 million tonnes of this waste is ending up in the ocean every year from the coastal zone or via rivers. It is also important to recognize that this is a growing problem. In the 1950s about 1.5 million tonnes of plastic was manufactured each year. By 2015 this figure had grown to 299 million tonnes. Low oil prices and poor future prospects for oil sales driven by decarbonization of the global economy are encouraging some oil-producing states to invest in growing their plastic manufacturing industry as an alternative use of oil. This, of course, will make the stuff even cheaper in the future.

I was aware of the plastic debris problem both from reading scientific papers and visits to the beach in the UK, the Canary Islands and other destinations where some level of trash was always present. However, it was the Thinking Deep expedition to Honduras in 2015, which I describe in the previous chapter, that really woke me up to the seriousness of the problem.

We had been diving for a couple of weeks around the island of Utila, which was always a pleasure compared to the waters of northern Europe as it was always beautifully clear. I could usually see over a distance of 30 metres or more underwater, giving a sense of enormous space in all dimensions. As we had been diving, though, it had been getting noticeably warmer and warmer. The air was heavy with heat and humidity as we set out one day to dive at the Maze on the other side of the island from the dive centre, a site of very confused reef near shore giving way to steep walls further out. The sky looked threatening with heavy cloud as we left and by the time we had moored to the diving buoy at the Maze the sky had turned black and the wind had begun to blow. There was the rumble of thunder and water spouts began to form around us and out to sea. The wind would drag a spiral of cloud down to the sea, forming a funnel, sometimes touching the water and other times dissipating and falling apart. I had never seen this phenomenon of nature before and I was deeply fascinated. There was a buzz of nerves among the divers – lightning can strike and kill a diver who is on or close to the surface – but we went ahead anyway as the storm had not yet arrived.

The dives went ahead as planned. As we headed back around the island the heavy rains of the tropical wet season arrived, sheets of large raindrops drenching everything in seconds. Over the following days it brought with it clouds of sandflies and mosquitoes. The former would swarm on our lips, in our eyes and on our faces as we tried to prepare our dive gear, incessantly biting and harassing us at all opportunities.

The following day was clear and we motored around to the Maze again. This time a very different scene greeted my eyes. As far as I could see, from the coast to the horizon, there was a giant floating raft of rubbish. Dirty plastic bottles, cartons, other types of food packaging, pallets, broken polystyrene, bits of vegetation, broken fibreglass, every piece of rubbish you could imagine lay around the boat. As I looked over the back of the dive boat I could even see plastic bags drifting past underwater onto the reef. We simply could not believe our eyes. Erika, our young Mexican diver, took photographs around the boat but these could not really convey the ghastly extent of the trash raft we found ourselves bobbing around in. A quiet depression settled over all of us, shrouded by a mixture of shock and somehow a shared shame at what humans were doing to the ocean. So thick was the disgusting slick of plastic on the surface that we even had to move it out of the way of the back of the boat to get our divers in the water.

'Where did this all come from?' I asked, at a complete loss as to how the ocean could be clear of debris one day, and the following we were faced with a picture from the apocalypse.

Erika replied that she had read about the Honduran government making a complaint to Guatemala about huge amounts of trash flowing down one of their rivers and being washed up on tourist beaches near the border.

When we returned to the dive centre I went in search of local papers and sure enough there were pictures of beaches with bulldozers trying to clear mounds of trash. A lack of waste management in several large towns on the river meant that their rubbish went into the river, and from there into the Gulf of Honduras. The local mayor across the border in Honduras was complaining to the government that the trash was ruining the local tourist industry. Who wants to go to a beach knee deep in foul plastic rubbish? I was astonished that the waste problem had reached such a serious level that it was causing a souring of relations between one country and another. The damage that was being done to the oceans was only just beginning to dawn on scientists and civil society.

When diving in Honduras over those weeks I sometimes encountered turtles, usually on the reef browsing on seaweed, or on sponges which grow on the reef in abundance. Seeing the raft of plastic arriving on the reef immediately brought these animals to mind as they are one of the most vulnerable to problems of entanglement and ingestion of human debris in the marine environment.

Many years ago, when my wife, Candida, and I went to the Maldives on honeymoon, we both got to see these animals up close for the first time. Turtles are said to look wise, which is not surprising as they live a long time, as long as humans or sometimes longer. They have a beaklike mouth, dark, intelligent eyes and a large head which resembles that of a bald sage. I remember the delight when we saw our first hawksbill turtle swimming past on the reef, the scaly skin a beautiful patchwork of greens and yellows and the shell a bewildering pattern of emeralds, browns, reds and yellows. They were surprisingly vigorous swimmers and would often come to check divers out, again that curiosity suggestive of intelligence.

At the time, the reefs of the Maldives Islands were yet to suffer from their first mass-bleaching event and they were spectacular. Clouds of brightly coloured damselfish and shoals of large sweetlips, parrotfish and other species danced among stunning gardens of corals, sea fans and sponges. White-tip and black-tip reef sharks were ubiquitous, the former looking like a smooth green, and much larger, version of the dogfish I saw in Ireland as a child. But a special treat for both of us was a dive on the reef at night. Candida and I were somewhat more experienced divers than most of the rest of the group on the dhoni, a long, flat-bottomed wooden boat used for diving in the islands. The dive master leading us signalled that we could go off and explore the reef ourselves ahead of the other divers, who were slow and rather clumsy in the water. This we did, a whole new world opening up to us through the tunnels of light formed by our torches. Unlike during the day time, the reef was now festooned with crinoids – sea lilies – appearing like great black flowers blooming from the coral heads. Lionfish pugnaciously spread their fins, beautiful in their white and red stripes. Moray eels foraged for prey investigating every hole and ledge for the unwary fish. Most remarkable, though, was a large green turtle, apparently just asleep resting on the reef. I floated, watching in amazement less than a metre away from it and yet the animal did not respond. It was definitely asleep and I could not believe that an air-breathing animal was able to stay underwater for so long. Little did I know then that green turtles can dive for as long as five hours without taking a breath. Their bodies are superbly adapted for storing oxygen in the blood and in the tissues making them one of the super divers of the aquatic air-breathing vertebrates.

There are seven species of marine turtles, and most of them have similar life cycles. They have to come on land to lay their eggs, which are buried in the sand at the top of nesting beaches.

Once the eggs have incubated, generally for about two months, they hatch at night as tiny versions of the adults and dig their way to the surface. The hatchlings then have to make a mad scramble to the sea to avoid being eaten by the many predators waiting above ground for them. Their enemies can vary from racoons to seabirds, such as gulls or pelicans depending on where they are in the world. This dash for safety is a scene familiar to many of us from watching natural history programmes on television. However, what happens thereafter are called the 'lost years'. During this time, the hatchlings, which in the case of a green turtle might be just 5 centimetres long, spend time in the open ocean. Much of what happens during this period remains a mystery, but scientists have gradually figured out that the young turtles feed on zooplankton, such as jellyfish, as they drift in ocean currents often around entire ocean basins. For green and logger-head turtles they may orbit the ocean for as few as three years or as long as 10 years. As they undergo this incredible journey many of them die, of course, eaten or succumbing to one of the many hazards of the ocean. Those that survive grow into juvenile turtles and eventually migrate back to coastal feeding grounds, generally close to where they hatched, with the exception of the leatherback turtle, the largest turtle at up to 700 kilograms in weight, which spends most of its life in the ocean only coming ashore to lay its eggs.

Depending on where they are and to what population they belong, even as adults the turtles can undergo staggering migrations between feeding grounds and their nesting beaches, to which the females consistently return and which are generally the same beaches where they themselves were hatched. One adult loggerhead turtle has been tracked from a nesting beach in Papua New Guinea to the west coast of America off Oregon, a distance of nearly 13,000 miles. These miraculous feats of navigation are at least partially achieved by detection of the Earth's magnetic field. Turtles have an inbuilt magnetic compass and while out in the ocean can keep themselves within the boundaries of currents by sensing the changes in the magnetic field. Adult turtles even memorize magnetic maps of their coastal feeding grounds. Such magnetic compasses are used by other animals, such as migrating birds.

Turtles as a group are incredibly ancient, their ancestors first evolving at the dawn of the age of the dinosaurs. They are a remarkably successful group of animals, being found throughout the tropical oceans and into temperate latitudes. They are, however, in serious trouble, with three species now classified as endangered or critically endangered and another three as vulnerable to extinction. Of course, it is human impact that is dragging them down the slippery sides of the extinction pit. While direct human predation and by-catch in fisheries are the two greatest threats to turtles, plastic debris is an increasing problem. The issue is that because plastic is everywhere in the ocean it is affecting turtles through every stage of their life history. This is all the worse because they are long-lived animals and take a long time to develop into adults. Nesting beaches are contaminated with plastic waste. Even before the turtles hatch, the presence of plastics in the beach alters the rate at which sand warms up and cools down and also its ability to retain moisture. The temperature of the nest determines the sex of the developing turtles so if the nest is cooler more male turtles may hatch than normal. The development rate of the embryos in the eggs will also be stunted and an increased permeability of the sand to moisture can lead to the eggs dehydrating. Once the hatchlings dig their way out of the nest, then they are vulnerable to becoming trapped or entangled in debris lying on the beach, especially lost fishing gear, such as nets or lines. The nesting females are similarly vulnerable to becoming enmeshed in trash washed up or discarded on beaches when they dig the holes into which they lay their eggs.

Out in the ocean, both as developing juveniles and as migrating adults, the turtles are faced with two dangers: ingestion of plastics or entanglement. Animals will gather under almost anything that floats in the ocean because it may provide shelter or be home to prey. Before humans, such objects might include floating seaweed or terrestrial vegetation like floating trees washed away from the land. Now much of what floats in the ocean is plastic. Turtles can ingest this plastic accidentally, for example in their food, especially when algae is growing on it, or deliberately, as in the case of pieces of white or clear plastic that might resemble jellyfish. This plastic can cause direct damage to the digestive system or intestinal blockage. As little as 0.5 grams of plastic and fibres have been found to be sufficient to block the digestive system of a juvenile green turtle and kill it. Alternatively, plastic can partially fill the stomach of a turtle, reducing its appetite or the capacity to feed, resulting in malnutrition, an affliction called dietary dilution. This can be so severe that the turtles can starve, especially young animals with small stomachs.

# Plastics and Other Pollutants

In some parts of the world 100% of stranded turtles have human trash in their guts, most of which is plastic, and it is estimated that around 50% of turtles overall have ingested such material. Some species are more vulnerable to the deadly effects of swallowed plastic than others. For example, loggerhead turtles have a very general diet and as a result have a wide intestinal tract. This means that they are better at getting rid of ingested plastic through defecation than other species. Entanglement of turtles is particularly associated with lost fishing gear such as nets. This causes a range of injuries to turtles, including maiming or amputation of limbs, increased drag, restricted movement, loss of buoyancy control and ultimately death. Some 5% of stranded turtles are entangled in nets or other human debris, of which more than 90% die. Not only is such a cause of mortality a tragedy for turtle populations, but the animal welfare aspects of entanglement are horrific. Turtles may take a very long time to die from the effects of being choked or otherwise injured by nets or other plastic garbage and no doubt suffer terribly in the process. Thinking about this, especially after seeing photographs of the way in which these animals die, really brings tears to my eyes.

Unfortunately, turtles often concentrate in areas such as oceanic fronts or convergence zones, where water masses of different properties meet. This is because such regions are often productive and concentrate the prey on which turtles feed. However, the physical properties of such regions mean they also accumulate trash. Such a phenomenon is known as an ecological trap where the migratory and feeding habitats of turtles place them in areas of the ocean with the highest concentrations of plastic debris, increasing the risks of ingestion or entanglement. Estimating the numbers of turtles killed by plastic debris each year is impossible given our current understanding of the problem. Many turtles killed by plastics will rot away at sea, sink or be eaten by scavengers, with only a small proportion washed up on shore. Expert opinion suggests that levels of mortality caused by trash is significant at the level of populations and is therefore contributing to the demise of these wonderful animals. Unfortunately, they are not the only ones.

Over 700 species of marine animals have been found to have eaten plastic or become entangled in it. Large marine predators, such as seabirds, whales and seals, are all vulnerable to plastic pollution. This is because they are long-lived with lifespans of tens or even up to 200 years in the case of the bowhead whale. This means they have a long time to pick up and accumulate plastic. They may also mistake plastic for food, happen to live or pass through areas with high levels of plastic pollution, or are inquisitive and get themselves tangled up in the process of investigating a strange object in the sea. For example, of the 21 known species of albatross, at least 12 have been found to have eaten plastic. The stunning wandering albatross, so familiar to me from my expeditions to the Southern Ocean, is one such species affected. These great white birds skim the waves of the Southern Ocean with narrow wings that may be up to 3.5 metres long, the longest of any bird. Wandering albatross feed by seizing prey at the ocean surface, mainly squid and fish. However, they are also voracious scavengers and will follow fishing vessels swallowing any discarded fish or other potential food in frenzied competition with each other. This generalist feeding strategy means that these great seabirds swallow plastic, but they seem to be able to regurgitate or pass it through their digestive systems with relatively little visible harm to the animal. Their nests may be littered with bits of fishing gear, mainly from longline fisheries as far away as the coast of South America.

In other species, such as the Laysan albatross which breeds mainly in the Hawaiian Islands and forages in the North Pacific, plastic ingestion by chicks is known to significantly reduce fledging weights, which is likely to cause a reduction in survival. The adults of this species feed through both scavenging and hunting of live prey, and again this generalist feeding strategy seems to lead to extremely high levels of plastic ingestion of everything from broken-up pieces of plastic to fishing gear. The occurrence of plastics in these birds is high, ranging from more than 80% of individuals assessed to 100%, depending on the study. The North Pacific is notorious for hosting the Great Pacific Garbage Patch. This was initially identified from an area between California and Hawaii where scientists deploying nets discovered a density of plastic debris of over 330,000 pieces weighing over 5 kilograms, per square kilometre. Although the number of pieces of plastic was outnumbered by plankton by about five times, the weight of plastic was about six times that of the plankton. The pieces of plastic vary from large remains of fishing nets, buoys and floats to tiny plastic fragments, pieces of sheeting and monofilament line.

The ocean currents across the North Pacific are complicated but can be pictured as two great parcels of water moving in a circular pattern, the sub-tropical gyre to the south and the sub-polar gyre to the north. The points where these gyres meet are known as convergence zones and here one water mass tends to slip beneath the other causing a downward movement of water. Areas where waters converge draw plastics towards them, and if the debris is sufficiently buoyant it floats above the area of down welling. This generates a narrow band of concentrated plastics lying across the Pacific, north of Hawaii. In the eastern and western parts of the sub-tropical gyre there are also smaller sub-gyres, semi-closed zones of recirculating water that also accumulate and retain plastic. This gives rise to three areas of debris accumulation, the one already mentioned, the Eastern Garbage Patch between Hawaii and California; another to the east of Japan, the Western Garbage Patch; and a third area of concentration in a band across the North Pacific. Populations of Laysan albatross that feed in the Western Garbage Patch can have up to 10 times the amount of plastic in their gut as birds that feed outside this zone. However, despite the evidence that Laysan albatrosses with high plastic burdens weigh less at fledging and the horrific images of dead chicks stuffed with plastic garbage familiar to us from nature documentaries, scientists are still unsure of the levels of mortality and harm done to the Laysan albatross by plastic ingestion. There simply haven't been enough studies to understand what is going on. While this is all extremely depressing, there is still another more insidious threat from plastics. Plastics themselves are formed by taking small molecules (monomers) and linking them together to form large molecules (polymers). Chemicals are used to initiate and promote this polymerization and then more chemicals are added to the plastic to give it specific properties related to its use. In some cases, the monomers themselves have toxic properties. Some of the chemicals used to promote polymerization are also highly toxic, especially to marine life. The additives include plasticizers, antioxidants, flame retardants and UV-stabilizers. Several of these have been identified as being potentially hazardous, such as phthalate plasticizers, which can make up 50% of some plastics, flame retardants and lead-based heat stabilizers. Other contaminants are generated as by-products of plastic manufacturing. Once the plastic is in the ocean it also begins to absorb and accumulate a complex mixture of other persistent chemical contaminants from seawater, mostly of human origin. Many of these chemicals are associated with toxic effects in marine life and in humans. The question is whether these chemicals, both those used in the manufacturing of plastics and those absorbed by plastics in seawater, can be transmitted to animals as a result of being eaten.

The big problem in trying to discern the influence of plastics in transferring harmful chemicals to marine life is that there are many other sources of contamination. An example is a group of chemicals that have been used as flame retardants in consumer electronics, cars, televisions, furniture, carpets and even in clothing known as poly-brominated diphenyl ethers (PBDEs). These were developed to replace another class of flame retardants, polychlorinated biphenyls (PCBs), because they were found to be highly toxic and persistent in the marine and terrestrial environment. PBDEs are highly stable chemicals that disperse as gases or on dust particles through the atmosphere. These chemicals enter the ocean and are then taken up by plankton. PBDEs are lipophilic, which means that they are absorbed and concentrated in the fats of animals. The plankton are eaten by larger animals, fish or crustaceans, and these in turn are fed on by larger predators and so on, the PBDEs becoming more and more concentrated in the fats of animals as they are moved up the food chain. This is called biomagnification or bioaccumulation. The chemicals are persistent, and they and their breakdown products remain in the environment and in the animals they contaminate for long periods of time. They are a particular problem for top predators in marine food chains which are not only exposed to a high level of PBDEs in their food, but also live a long time so that these chemicals accumulate in their body fats.

Topping the marine predator chain is the killer whale. I have seen these wolves of the sea cruising up and down fjords of South Georgia. They are breathtaking animals, and to see their smooth black, shiny backs curving through the water topped with that iconic tall fin is one of nature's true spectacles. They move with purpose, hunting for young seals taking their first tentative dip in the choppy waters around the island. The more we learn about these animals, the more complex their society and behaviour is revealed to be. In Antarctica there are at least four main types of killer whales, which differ in their size and in their colour patterns, particularly the position and size of the white patch above and behind the eye.

Type A are the largest with the male growing up to 9 metres long. They typically live in open water where they hunt minke whales.

# Plastics and Other Pollutants

Type B typically live inshore and in among the pack ice where they specialize in hunting seals but will also take minke whales and humpback whales should the opportunity arise.

Type C has been observed mainly in east Antarctica in areas like the Ross Sea and feed on toothfish, a large predatory fish that lives near the seabed and which is a prized target of longline fishing in the Antarctic.

Type D has an extremely small white eyepatch and lives in the sub-Antarctic, existing all around the continent. Little is known about the diet of this form of killer whale but it has been observed to steal toothfish from the longlines of fishing vessels around French Antarctic Territory, a habit shared by killer whales and sperm whales elsewhere in the Antarctic.

These different forms of killer whales are genetically different and some scientists think they may be different species.

Off the coast of Canada there are two distinct types of killer whale, different from those in the Antarctic: those that are resident and live inshore and specialize in eating fish, and those that are transient and specialize more in eating seals and other whales. In 2006 it was found that in both of these types levels of PCBs, the legacy pollutant, and PBDEs, the replacement flame retardant, had reached extraordinary levels in both resident and transient killer whales. In the resident killer whales levels had reached around 150 milligrams per kilogram of fat for PCBs and just below 1,000 micrograms per kilogram of fat for PBDEs. For the transient killer whales, PCBs had reached 250 milligrams per kilogram of fat and PBDEs 1,000 micrograms per kilogram of fat. The reason for this was that the same chemicals were in their prey, mainly salmon for the inshore type and seals for the offshore. The threshold for toxic effects from PCBs was exceeded by more than 10 times in these killer whales.

Toxicity thresholds are less understood for PBDEs, but both chemicals disrupt vitamin-A levels and interfere with the function of the thyroid leading to impacts on development of the brain, the nervous system, reproduction and the immune system. PBDEs may also mimic hormones in the body, such as oestrogen. To make matters worse, these chemicals are transferred to killer whale calves via their mothers' milk, which is rich in fat. When such high levels of these pollutants were discovered in Canadian killer whales, the phrase 'fireproof whales' was coined. The levels of pollutants were so high in these animals that the resultant toxic effects were almost certainly a threat to their continued existence. However, because there are several routes for these flame retardants into the ocean, there is little understanding of the role of plastics in transferring them to marine animals.

Most people will be aware of the hazards that oil presents to marine life, particularly seabirds. You may also be aware of problems associated with heavy metals, such as mercury, which like PCBs and PBDEs can accumulate up the food chain. Fish such as tuna or swordfish naturally accumulate such metals in their bodies. Human pollution of the ocean adds to this burden, and this is why eating too much of these large predatory fish is not a good idea. Many of these pollution problems have been around for decades and are well studied. Plastics are a relatively new problem that, thanks to *Blue Planet II* and the activities of many NGOs and campaigners, is now very much in the public eye. However, as I've pointed out, while common sense tells us the presence of plastic in the marine environment is not a good thing, proving that it is having a significant and detrimental impact on marine life and ecosystems is not easy.

Consider the cocktail of chemicals that you are exposed to every day through the use of pharmaceuticals and common products in the home, many of which end up in the ocean.

A few years ago, I remember standing on a sheltered beach in Lanzarote known as Playa Chica. I had been training all day for my dive master qualification and had just walked through the sand feeling exhausted. As the holidaymakers gathered their towels and beach bags I glanced out to sea. Playa Chica is a small embayment with a quay on one side and rocks on the other. All the dive schools use it for training and as I looked at other divers surfacing I noticed the sea surface was unnaturally smooth and oily-looking. It was a slick of sunscreen, washed off swimmers and paddlers of all ages into the beautiful and clear waters of the Canary Islands. I had been diving there for over a week and had seen marine life that was long gone from the Mediterranean. Angel sharks, flattened 2-metre-long mottled-grey monsters that lurked on the seabed waiting to ambush prey; sea hares like the ones I saw as a child in Ireland, but the size and shape of a rugby ball; large groupers and clouds of damselfish hanging off rocky reefs, which themselves were home to brightly coloured sponges, patches of sea squirts that resembled baked beans, and bright yellow corals. The Canary Islands, judging by the crowds of divers getting changed in the car park by Playa Chica, was becoming a popular dive spot for Europeans. They were coming because of the abundant marine life around islands like Lanzarote. The almost invisible effect of so many visitors was there staring me in the face as the sun dropped towards the horizon, turning the sky orange and highlighting the slick of oily sunscreen on the surface of the once crystal-blue waters. In that moment it was clear to me that such large quantities of sunscreen containing oxybenzone and other chemical additives must be having an effect on the marine life in the coastal waters of the Canaries.

Oxybenzone is used in pretty much every cosmetic or domestic product you can think of: sunscreen, body fragrances, hair-styling products, shampoo, conditioners, anti-ageing creams, lip balms, mascara, insect repellent, dishwasher soaps, washing-up liquids, hand soaps, bath oils and salts. It has even been used as a food additive. In fact, when I checked my shower gel this morning, to my horror, there it was in the list of ingredients as benzophenone-3. It is used in sunscreens to block UV and likewise in other products to protect the other chemicals from being changed by sunlight. It has been recognized as an emerging environmental contaminant of concern by the US Environmental Protection Agency. About 6,000–14,000 tonnes of sunscreen, containing 1%–10% oxybenzone, are estimated to be released on coral reefs each year.

It has been proven that levels of oxybenzone found in seawater in popular tourist beaches surrounded by coral reef in areas of the Caribbean and Hawaii are toxic to corals. Coral larvae resemble tiny slugs covered in minute hairs called cilia, which beat and propel the animal through the water. When exposed to oxybenzone the larvae become deformed, stop swimming and die. Microscopic examination shows that oxybenzone causes the cells of the larvae to die, they also lose their symbiotic algae, the zooxanthellae, bleaching like adult corals when under stress, and they become encased in a shell of calcium carbonate. The DNA of the larvae is also damaged. Oxybenzone has also been found to be toxic to crustaceans and fish, and in the latter sublethal concentrations have been shown to have effects on reproduction and development.

Humans are also exposed to oxybenzone. Some of it is absorbed through the skin from the products we use but it also gets into the body through drinking water and food. Recent studies on humans indicate that oxybenzone is able to cross the blood-brain barrier and can be detected in the hypothalamus and white matter of the brain. What this means for us humans is largely unknown, although we know that the hypothalamus is extremely important in receiving hormonal signals to the brain and controlling various processes in the body. There is evidence that high oxybenzone exposure in women leads to a reduction in the length of pregnancy and also influences the birth weight of babies, dependent on their sex (increase in birth weight of boys but a decrease in girls). The effects are of a similar level to those associated with organophosphorus pesticide exposure – chemicals that are known to be highly toxic to humans. It has also been implicated in the occurrence of Hirschsprung's disease, a developmental abnormality of the human intestine that causes it to become blocked. In rats oxybenzone crosses the placental barrier and damages the brain of embryos potentially leading to neurological disorders after they are born.

The Deep

I don't think I need to labour the point any further. This chemical, along with many others used in personal-care products, is a problem to the environment and a problem for us. Oxybenzone is an endocrine disruptor. This means it either mimics or influences the effects of hormones in the body. These are biochemicals, such as testosterone or oestrogen, that even in tiny quantities act as messengers in the body commanding critical processes in development and reproduction. Disruption of the functions of such biochemical messengers have far-reaching consequences for animals and us humans. Many of the chemicals I have discussed in this chapter, including phthalates, PCBs and PBDEs, are also implicated as endocrine disruptors. Oxybenzone has also been classed as pseudo-persistent in the environment. In other words, we constantly top-up the quantities present through its use in everyday household products. There is evidence, though, that it may also be biomagnified up the food chain like PBDEs, although not as efficiently.

This is just one substance among thousands that are used by us every day that eventually end up in the ocean and potentially cause harm.

So what has been done about such an insidious problem? Certainly where there is incontrovertible evidence that such chemicals are poisonous to the environment and to humans, steps have been made to ban them, usually through a phaseout of their use. The international agreement under which such chemicals are banned is called the Stockholm Convention, which was brought about to prevent the use of damaging pesticides such as DDT. PCBs are a good example. These chemicals have been banned from use in many countries and were replaced by PBDEs. When the PBDEs were found to be harmful these were banned under the Stockholm Convention, starting with the most toxic of the group. PBDEs are now being replaced by novel flame retardants (NFRs), which include a range of brominated flame retardants and organophosphate flame retardants. Like PBDEs these new flame retardants occur in our homes and will find their way into the environment. I wonder how long it will be before these chemicals will also be found to be harmful to marine life and perhaps also to us.

Two days before writing these very words I received a jubilant email from Craig Downs, a larger-than-life scientist of Hawaiian descent who studies the impact of pollutants on coral reefs. Hawaii had just banned sunscreens containing oxybenzone and octinoxate, the latter being another chemical associated with toxic effects on marine life and showing endocrine disrupting activities. This was largely because of the scientific work that he and his colleagues had undertaken demonstrating the toxicity of these compounds on corals. As I sat at my computer in Oxfordshire I could imagine Craig, who is incredibly passionate about his work, dancing around his office in joy. Craig sent me links to the headlines flashing all around the world on this extraordinary move banning a substance many holidaymakers in Hawaii used every day. Some of the news articles also held comment from the cosmetics industry, which condemned the move as putting people at risk from the harmful effects of ultraviolet (UV) radiation. Given the dramatic rise of skin cancers in countries like the USA, and having had a carcinoma removed from my forehead only last year, I am well aware of the importance of protecting myself and my family from damaging UV radiation. Craig, however, had gone to great lengths to point out that alternative chemicals which actually give protection across a broader range of UV wavelengths, such as zinc oxide, are available. These alternatives are much less damaging to corals and other marine animals, and probably also less of a risk to us. There are other measures that can be taken to protect ourselves against the sun too, such as wearing UV-protective clothing or staying in the shade. Craig's victory and the response of the cosmetics and chemical industry immediately reminded me of one of my marine biology heroes from the past, Rachel Carson.

Rachel was a marine biologist living in the United States who also had a gift for writing. In the 1950s she had published several successful books on the ocean, including *The Sea Around Us* and *The Edge of the Sea*. In the late 1950s to early 1960s, Rachel turned her attention to a rising environmental crisis: the use of pesticides in modern agriculture in the United States and elsewhere in the developed world. In places like Britain, this crisis manifested itself through a number of sinister events. In 1959, a mass death of foxes occurred in and around Oundle in Northamptonshire, and in the spring of 1961, tens of thousands of birds were found dead or dying in agony in the English countryside. Birds of prey like the peregrine falcon all but disappeared. The culprit was the indiscriminate use of a range of pesticides, such as mercury, benzene hexachloride, dieldrin and DDT, the latter two organophosphorus pesticides. These chemicals were applied as seed dressings or sprayed over crops. Rachel accrued the evidence of the harm these chemicals were doing to the environment, and in 1962, following a delay in publication as a result of poor health related to breast cancer, she published the book *Silent Spring*. The book focused on the damaging effects of the indiscriminate and liberal use of pesticides on wildlife, and touched on the toxicity of these chemicals to humans. *Silent Spring* was also highly critical of the chemical industry for spreading disinformation regarding their products and public officials of accepting their claims without question. I can only admire from afar the bravery of this woman who was, by all accounts, a shy and reclusive character. *Silent Spring* and its author were attacked savagely by the chemical industry and even by scientists who supported the use of such chemicals to boost crop yields. Rachel was attacked personally as not having the expertise to make claims regarding the use of such chemicals as she was a marine biologist with no training in biochemistry. An ex-US Secretary of Agriculture even wrote a letter to the ex-President of the USA, Dwight D. Eisenhower, suggesting Rachel was a communist. However, it came to light that *Silent Spring* had been reviewed by many experts in the field, and a large body of scientists supported the findings of the book. The book led to the banning of the indiscriminate use of DDT, the eventual creation of the United States Environmental Protection Agency and was influential in the eventual establishment of the Stockholm Convention. It has since been recognized as one of the most significant contributions to Western literature and contributed to the emergence of modern environmentalism.

The final chapter of Rachel's book begins with a remarkable passage that echoes through to the present day with respect to the ongoing use of chemicals that harm the ocean. She likens the situation in the early 1960s to choosing between two roads. The first is a deceptively easy road of rapid progression in industrialized agriculture, including all that is implied by the use of pesticides and other chemicals. The second choice is a more difficult road, but one that may offer the chance of feeding humanity and also preserving the environment. She makes the point that we have a right to know about the nature of the chemicals that are being used to produce our food and also to decide not to accept their use if they are shown to be harmful to the environment. Instead, we should look at other solutions to the issue of securing our supply of food.

I will return to the sense of the choice between the two roads in the last chapter of this book. In the past, the toxicities of chemicals were assessed by dosing animals with increasing concentrations until a certain number died. Such an approach is completely inadequate to really understand the impact of chemicals on the physiology of living organisms at environmentally realistic concentrations, after they have been diluted in seawater. We have now developed the technology to look for the specific markers of disruption of processes at the level of cells and organs within microbes, animals, algae and plants living in the wild. Such methods give a better view of the modes of action of chemicals in the environment, and by measuring parameters such as growth, reproductive output and changes in behaviour we can assess the overall impact on species, communities of organisms and ultimately on ecosystem function.

So why are we in a situation where substances like plastics, their additives and other chemicals are almost everywhere in the ocean? One reason is that somehow it has become acceptable to use the ocean as a giant dustbin. It costs manufacturers very little to have their products flushed into the sewage system or discarded into the environment. Effectively, they are using the ocean as a free garbage-disposal facility. They are not held accountable for the long-term costs to the environment or ultimately to the whole of society, including future generations. What about us as individuals? Many of us care about the environment and wildlife and would never consciously harm the ocean. However, we have simply come to accept the use of plastics and chemicals in almost every facet of our lives in the assumption that they are safe. How many of us look at the ingredients of personal-care products like shower gels, sunscreens, nail varnish, lip gloss, deodorant, perfumes? Even if we do can we even understand what the long list of chemicals means?

A number of things clearly need to happen. Firstly, the only way to deal with the plastics problem is to find more environmentally friendly alternative materials, the philosophy put forward by Rachel Carson. Replacement of single-use plastics will take time, so an immediate priority must be to stop the stuff getting into the ocean. More effective solid waste management is critical, especially in resource-poor countries where it is almost non-existent. Here, the corporations that manufacture and use plastics have a responsibility to consider the entire life cycle of their products, and where possible to ensure it is recycled or disposed of properly.

The answer to the burgeoning problem of complex chemicals

in everything from pharmaceuticals to shampoo is complex, but overall a new strategy is required. For those materials we just cannot do without as a society, such as medicines, more effective waste-water treatment is needed. But for many chemicals, for example those that are used as flame retardants, cosmetics, other forms of personal-care products, and whatever other uses are out there, we need to ask the question as to whether we really need these chemicals and ask if there are alternatives. This was the very approach that Rachel Carson promoted. Here the principle of reversal of the burden of proof must be considered. If you are a chemical company or a corporation making such products, there should be a legal obligation to prove that what is being sold to the public has an acceptably low level of harm to the environment given its use. This includes harm to both land-based ecosystems and the ocean. Part of the reason that such chemicals are appearing in such profusion is that taxpayers cannot bear the burden of such extensive testing. This therefore has to fall on those who are profiting from the synthesis of such materials. After all, if they are using the ocean as a cheap waste disposal service, they should be obliged to prove that their activity is doing minimal harm.

As *Silent Spring* so eloquently made clear, we are on the road to ruin if this problem cannot be solved. These are tough choices for society to make and no doubt will be strongly opposed by the companies that make all the products of modern convenience and governments as well, as we have discovered time and again over the last century in the cases of pesticides to cigarettes. For the individual, of course, this can seem very daunting.

# Plastics and Other Pollutants

A friend of mine, Nigel Winser, the director of Earthwatch and long-time environmentalist, once said to me: 'You and I as members of society have two sources of power: the power to vote and the power to decide on what we buy.' These words have stayed with me ever since. I am now well aware that we should not buy single-use plastics if we can possibly avoid them. It has become relatively easy to replace plastic cups and water bottles with reusable containers for our drinks. I always urge friends of mine to look at the ingredients list of the shampoo, shower gel or perfume that they are buying. If you don't recognize any of the chemicals on the list of ingredients, look them up on your phone, tablet or computer. You might be unpleasantly surprised by what you find and then making that informed consumer choice will be easy.

These are small but vital choices. Choices that are likely to benefit our own health as well as the health of the ocean – and we *all* have a part to play in making these informed decisions as individuals. I think about the horrible suffering caused by plastic waste wounding, choking or starving animals like turtles to death. I also think about some of the most majestic and spectacular animals of the ocean, like the killer whale, accumulating the poisons we are releasing into the ocean, and dwindling and winking out in the future as a result. But this is also a story about our own future. As human populations grow we are becoming increasingly reliant on the ocean for our food, whether it is wild fish or farmed seafood. Are we really going to accept increasing contamination of our food sources with plastic and noxious chemicals? Some might say it's a no-brainer!

# 7

# The Changing Ocean:
## *Running Out of Air*

It may surprise you to know that while you have been reading this book, every second breath you have taken comes from the ocean. The single-celled algae that make up phytoplankton floating in the upper sunlit layers of the ocean photosynthesize – just as plants do on the land – and a by-product of this process is the generation of oxygen. This oxygen is released from the ocean and helps to maintain our atmosphere. As phytoplankton cells and other organic material sink from the ocean's surface, forming marine snow, they are broken down by bacteria, which use oxygen. The result of this is a dip in oxygen concentrations throughout the ocean between the surface and depths of 1,000 to 1,500 metres. This is called the oxygen minimum zone or layer. In deeper waters oxygen is replenished by deep, cold, oxygenated currents coming from the polar regions. At times in the past, patterns of circulation have broken down leading to reduced oxygen (hypoxia) or even an absence of oxygen (anoxia) across large areas of the deep ocean. Such disturbances to the availability of oxygen in ancient seas have resulted in phases of extinction of species of marine life.

It is not widely appreciated that disturbance to the natural oxygenation of the ocean is a signature of human impacts on it in the present day. The careless use of agricultural fertilizers and the effects of climate change on ocean temperature and mixing both reduce oxygen concentrations in marine ecosystems. These changes are occurring incredibly quickly by geological standards. There are many sayings related to the folly of ignoring history, and this is as true of our natural history as it is of the history of our civilizations. Like a sinister omen left behind in ancient ruins, rocks and fossils tell us that we should be very concerned about any human activities that disturb the distribution of oxygen in the ocean. It has the potential to affect all marine life, from microbes to iconic predators such as tuna and sharks.

In 1994 I flew out to Oman to join the RRS *Discovery* Cruise 211 to investigate the oxygen minimum zone in the north-west Indian Ocean. In this part of the world, highly productive surface waters combined with sluggish ocean circulation below, lead to extreme hypoxia. The rich rain of organic material, dead algal cells, the bodies of tiny creatures and faecal pellets from the surface feed bacteria in the water column to the point where they consume most of the oxygen in the water. The northern Arabian Sea, from Oman, around the coast of Pakistan to India thus forms an extreme version of the oxygen minimum zone, called an oxygen-depleted zone. Such zones exist naturally in several regions of the global ocean where high surface algal growth combines with poor water circulation so that oxygen is not renewed and depletion occurs. These include the Bay of Bengal, the western margins of North and South America and areas off the western coast of Africa, including South Africa and Mauritania.

For me the trip to the Indian Ocean was immensely exciting as it was only my second major ocean expedition. Oman seemed a very exotic location – extremely hot, with clear blue skies and a souk near to the port of Muscat in which we scientists foraged for gifts for those back home. Ahead of the trip the mood among the group was light-hearted, and I was eager to get going. On board *Discovery* I found myself in a cabin next to one of the grandest names in marine science, Amélie Scheltema of the world-renowned Woods Hole Oceanographic Institution, Massachusetts, who studied a particular group of wormlike molluscs called the Aplacophora. The animals are small, usually less than a centimetre long, and often covered in fine golden spines. We immediately got on very well. Amélie was in her mid-60s at this point, and as well as bringing with her decades of knowledge, she read copious quantities of science fiction. As a fellow fan, we swapped books during the trip. Also present on the ship was Lisa Levin, from the Scripps Institute of Oceanography and another of my marine biology heroes. As a young scientist it was a little difficult to approach such internationally renowned personalities in marine biology, but such barriers usually broke down when people were crowded together on a ship. John Gage, who I knew from the Marine Biological Association where I had worked in Plymouth, led the cruise and seemed in his element rushing around energetically and organizing things. My friend Paul Tyler acted as his second-in-command on this expedition. He took charge of sampling for the large animals that lived on the seabed so we inevitably ended up spending a lot of time working together over the next month. As usual Paul's unofficial role was to provide a steady diplomatic hand in resolving any difficulties that arose among the many 'alpha' scientists on board.

Lisa wanted to do a lot of box coring of the sediment as she wanted to see how the abundance and biomass of the small animals living on and in the fine mud changed with different oxygen concentrations. Paul and I wanted the bigger animals, such as crabs, sea urchins and sea cucumbers, and for these trawling the soft muds of the deep seabed was the mainstay of sampling at the time. To do this we used a very primitive device called an Agassiz trawl, named after the famous nineteenth-century oceanographer Alexander Agassiz, who worked on the sea urchins from the *Challenger* expedition as well as on coral reefs. The voyage of the *Challenger* between 1872 and 1876 is widely regarded as the world's first modern global oceanographic expedition. The Agassiz trawl comprised a simple metal frame with a net attached.

One way of fitting in all the sampling requirements of the scientists on board the ship was to work 24 hours a day and to split the time up across the various teams. Box coring was a fairly tricky operation, and the samples needed dividing among various different scientists, so it was decided to do this mainly in daylight hours. The trawlers, led by Paul, and including me, were mainly consigned to night-time sampling and we rapidly acquired the title of the 'night shift'.

As we steamed down the coast of Oman and past the island of Masirah under brilliant turquoise skies on a royal-blue ocean the excitement mounted. We had no idea what we might find in the oxygen-depleted zone. A Royal Navy Lynx helicopter appeared out of nowhere and buzzed the ship, perhaps in acknowledgement of a British vessel in distant waters, and the heat of the day began to climb. Very soon, the black-painted sides of *Discovery* became so hot that it was impossible to touch them. Flying fish, and even the odd flying squid, an animal I did not even know existed until that point, sprang from the path of the ship, curving away to port and starboard. We could see crabs too, swimming on the surface of the ocean. We began sampling the neuston – the animals that live on the interface between the air and the ocean's surface – using the rather unsophisticated device of a bucket, so we could study them alive on the ship. What we found were clearly swimming crabs, a species called *Charybdis smithii*, adapted to spend their lives in the upper layers of the western Indian Ocean and Arabian Sea. They had narrow pincers and their rear legs were paddle-shaped to assist with swimming. The crabs grew to a maximum size of about 7 centimetres across the shell and could occur in vast swarms, mainly coming to the surface at night. They could make up 90% of the biomass of small swimming animals in the Arabian Sea and were a prime food for tuna and other large fish. These crabs weren't the only surprising animals to show up in our buckets; *Halobates*, a fantastically blue insect, which resembled a pond skater, also made an appearance.

Before then, I had had no idea there was such a thing as a fully marine insect. The Arabian Sea was a truly strange place compared to everywhere else I'd been, in my thus limited experience.

The 'night shift' began to trawl the oxygen-depleted zone which started within 100 metres of the surface, reaching a minimum of oxygen at around 400 metres depth and then climbing again to depths of about 1,200 metres. In the core of the oxygen-depleted zone, the trawl came up laden with foul-smelling black mud, the typical rotten-egg stench of hydrogen sulphide. However, large numbers of strange animals were also recovered. The most distinctive of these was a small spider crab, *Encephaloides armstrongi*. These crustaceans were a rich orangey red, with a small body, only a couple of centimetres across, covered in small conical bumps, and with long spindly legs probably three or four times longer than the body width. The most notable features of these mini spider crabs were greatly inflated gill chambers, which appeared as distinct bulges on either side of the rear of the crab.

As scientists more than a hundred years before me had guessed, it was clear to see from looking at the adaptation of the gill apparatus of *Encephaloides* and other crustaceans that oxygen was in short supply on the continental slope of India, in the Arabian Sea and Bay of Bengal.

We began to recover the crabs in large numbers in some trawls, and our trawling rapidly produced other peculiarities. One day we were working in the morning rather than at night, on the slope off Oman, south of Masirah Island. We landed the trawl and began the arduous task of shovelling a very large load of mud off the deck in searing temperatures. Pretty soon my T-shirt was soaked in sweat, but as I was shovelling I spotted a spherical shape in the mud. I picked the slimy object out from the ooze to find a ball of jelly the size and appearance of a large glass marble.

'Paul, what's this?' I asked, offering the strange object to him to examine.

He was as baffled as I was. 'It might be one of Andy's beasts! Let me get him.'

Our colleague Andrew Gooday was a quietly spoken world expert in foraminifera, a group of amoeba-like single-celled animals that can be immensely abundant in the ocean. Forams can live in the plankton or on the seabed, are often tiny and have microscopic calcareous shells that can be elaborately shaped and sculpted. Some, however, can get to larger sizes. When Andy joined us he immediately became very animated on seeing the object in my hand. 'Yes, I think it is a large naked foraminiferan. I've never seen one so large or with this spherical shape.'

Andy took his prize away to the laboratory and we collected more of them from the mud, dropping them into a bucket. It later turned out to be a new species, *Gromia sphaerica*, from a group of amoebas never before seen in the deep sea. The jelly ball amoeba was also one of the most abundant large animals at depths of around 1,600 metres below the core of the oxygen-depleted zone. It was likely to be important in the seafloor ecosystem at these depths by virtue of its numbers. Similar animals had been seen at higher densities in shallower waters around the Antarctic.

Later that afternoon, some of the younger scientists were begging the crew to go swimming. The bosun grunted: 'No chance the old man will let you do that. The Knobby Clarks are out there. You don't see 'em but they're following the ship, guarantee it.' 'Old man' referred to the ship's captain and 'Knobby Clarks' was rhyming slang for sharks, of course. Sure enough that evening as we were standing on the afterdeck watching the sun go down, a large dorsal fin arose in the wake of the ship and slowly, sinisterly, sank out of sight. No one asked to go swimming again.

Over the next few days a large contraption was constructed on the back deck. It comprised a thick steel frame painted bright yellow with various instruments attached to it. This was a lander system for measuring the chemistry of the sediments in the oxygen minimum zone. There was to be a test where the entire instrument, which was nearly as tall as a person and a couple of metres across, was lowered into the water. We all watched as it was hoisted up by the A-frame at the back of the ship where the crew were ready with ropes to steady it. The lander was slowly dropped into the sea and bobbed around. John Gage was on the radio to the bridge directing operations enthusiastically from the aft deck. 'OK, bridge, can you just back the ship up a few metres and we'll pick it up?'

'What do you think this is, an effing London bus?' came the sharp reply from an unidentified officer on the bridge, much to the hilarity of all standing around. John didn't quite know how to respond, but by then the ship had moved into position for recovery of the lander. Clearly things were getting a little tetchy between the Chief Scientist and the officers of the watch.

The following day the lander was deployed for the first time to undertake its measurements. It was a bright, calm, sunny day so no problems with recovery were anticipated. However, when it was time to retrieve the lander the wind had blown up and the sea had become quite choppy. The acoustic signal was sent by the ship and the indications were that the lander was on its way back to the surface. The scientists stood scanning the sea with binoculars but the waves made it difficult to spot anything at the surface level. This valuable piece of equipment was very low to the surface of the water without any form of signalling apparatus on board. Now I understood why the crew had jokily been bidding their farewells when the instrument went into the water. It was never seen again. £80,000 of equipment simply drifted off into the Arabian Sea. It was my first experience of the things that could go wrong when carrying out deep-ocean science. Once something went in the water there was no guarantee of getting it back.

It wasn't all bad news on this expedition, though. One piece of equipment that was deployed successfully was WASP: Wide-Angle Seabed Photography. This was an earlier version of the towed camera system that we were to use when searching for the vents in the Southern Ocean. The vehicle was 'flown' above the seabed using altimeters and a depth-tracking system and the photographs recovered when WASP was brought back on deck. The images that were recovered were astonishing. At depths of 350 metres the seabed was covered in the *Encephaloides armstrongi* spider crabs, all marching in the same direction. In the core of the oxygen minimum zone, large animals were virtually absent. Below the oxygen minimum zone, where oxygen concentrations were recovering to near normal, the jelly balls appeared. They were clearly visible as shiny bubbles dotted in small clusters partially embedded in the sediment. Around each jelly ball there was a halo of paler sediment where the animals had been feeding on organic material from the surface of the seabed. It was amazing to me that these were a single giant cell.

In a canyon below the oxygen minimum zone we found an entire lake formed of dead jellyfish. They lay on the seabed as a patchwork of green and blue decaying bodies, a newly discovered rich source of food for animals in the deep sea. Elsewhere we found that these jellyfish had decayed to a centimetre-thick layer of gelatinous slime at the bottom of the ocean. Along with previous discoveries that the swimming crabs we had seen at the ocean surface could die in large numbers and form mass food falls in the Arabian Sea, this indicated a new route for the transport of organic carbon from the surface into the deep sea.

Five weeks later, I returned to the UK and the MBA with samples of the spider crabs and some squat lobsters and I began to work on the connectivity of populations across the oxygen minimum zone. What I had seen in the extreme oxygen-depleted zone off Oman intrigued me. Very few large animals lived within the low-oxygen zone. Those that were present just above or below the oxygen minimum were animals that had specially adapted to low-oxygen conditions. As well as *Encephaloides* with its inflated gill chambers, we found a tiny pearlescent mussel which lived in a nest of threads it secreted, presumably to regulate its environment in some way. Mats of bacteria were found on the surface of the sediment living by sulphide oxidation, a form of chemosynthesis similar to that on deep-sea hydrothermal vents. Smaller animals, such as worms, were found right through the extreme low-oxygen zone but showed adaptations, such as large gills, and some lived in balls of mud on the surface of the seabed. As expected, many groups of marine animals, such as urchins and starfish, were absent, and snails and bivalves were relatively rare in the lowest oxygen conditions. We knew that some of these groups had a low tolerance of hypoxic waters. Could such low-oxygen conditions affect evolution in the deep sea by acting as a barrier to animals that live in normal oxygen conditions? If so, they might isolate shallow-water populations from deep-water ones or isolate populations on either side of a low-oxygen zone. Isolation was a prerequisite for populations to evolve into new species in the ocean. Could animals 'hide' in low-oxygen zones, escaping from their predators that needed high concentrations of oxygen? Might they adapt to low-oxygen conditions to take advantage of the extra food present in these zones because of there being fewer animals to eat it?

As I began to delve into scientific papers on this topic I came across fossil evidence of the effects of low oxygen levels in oceans of the past. What I found was startling, and I felt it had profound implications for the impacts of global climate change today. Low-oxygen conditions produce assemblages of fossil animals that were low in species numbers. As in the oxygen minimum zone off Oman, relatively few species of animals could survive in low-oxygen conditions in oceans of the past. There were also specific chemical signatures of low-oxygen conditions in rocks, such as the presence of high amounts of iron pyrites (fool's gold) or high amounts of organic carbon, such as in black shales.

# The Deep

Geological evidence suggests that the Earth has alternated many times between warm periods of the climate and cooler periods. During the cooler periods, polar ice caps existed and cold, dense, richly oxygenated seawater sank and spread out across the deep sea. At lower latitudes deep water upwells to the surface, warming up and being circulated back towards the poles. This is known as the thermohaline circulation and is how the ocean works today, oxygenating the deep sea and spreading heat across the world, maintaining climate at temperatures that are comfortable for life as we know it. However, at other times in the past, possibly linked to natural phenomena such as volcanism or even to biological activity, $CO_2$ levels became high in the atmosphere leading to high temperatures and a breakdown of the thermohaline circulation. Ocean warming would have also led to a phenomenon known as increased stratification where a layer of warm water lies over cooler waters below, reducing mixing between shallow and deeper waters. Added to this, warmer water contains less oxygen. The overall effect of these changes was to reduce oxygen levels in the deep sea leading to widespread hypoxia (low oxygen) or even anoxia (no oxygen). The impact on deep-sea life was devastating. Tracing back through time there is evidence for these episodic events all the way back to the geological period called the Cambrian, more than 500 million years ago. Ocean hypoxia or anoxia often coincided with the extinction of large numbers of species of marine animals, for example ammonites, squid-like animals with coiled shells that swam through ancient seas. A living relative, the nautilus, lives on tropical reefs today.

The granddaddy of all extinctions in the ocean and on land,

much worse than the extinction of the dinosaurs, came at the boundary of the Permian and Triassic periods, about 251 million years ago. This extinction killed more than 90% of all marine species and has become known as the 'Great Dying'. Try to imagine what such an extinction might look like on the land or in the sea now. Nine out of 10 of the birds you are familiar with disappearing, or nine out of 10 of the common fish species around your coasts dying off. Prior to the Great Dying there were rich coral reefs, in some cases greater in size than the Great Barrier Reef before the recent mass-bleaching events. The reefs were formed by different groups of corals from those found today, the rugose and tabulate corals, some of which I had found fossils of as a child in Ireland. Animals such as brachiopods, that resembled clams in appearance, stalked crinoids or sea lilies, only seen in the deep sea today, and bryozoans, which formed miniature treelike or bush-like colonies grew in profusion in shallow seas. These entire ecosystems were wiped out in the environmental cataclysm that descended upon Earth. Whole groups of animals, such as the rugose and tabulate corals and the trilobites, multi-segmented creatures from the same group as insects and crustaceans, went extinct after existing on Earth, in some cases, for hundreds of millions of years. The coral reefs of the Palaeozoic era were obliterated, and nothing resembling a coral reef reappears in the fossil record for another eight to nine million years. On land, plant life was equally devastated and the flora was replaced by one dominated by ferns and lycopods, both spore-forming plants – in other words, plants capable of surviving harsh conditions. The vegetation die-off on land allowed massive flooding and erosion to occur. Seventy per cent of the terrestrial vertebrates also died out. Following the event, the dinosaurs rose to dominate the terrestrial megafauna.

Palaeontologists have long argued over the actual cause of this event and whether there was a single episode of extinction or several. Various causes for the extinction(s) have been put forward from a meteorite strike, like the one that killed off the dinosaurs, to a glaciation that caused sea levels to drop, destroying shallow-water habitats like coral reefs. Numerous studies and the discovery of new fossil evidence now suggest that the extinctions occurred over a relatively short time, a geological blink of the eye. The trigger appears to have been a massive outpouring of lava forming the Siberian lava traps. The remains of this flood of lava are now present in northern Russia as a massive volume of basalt rock up to 3.5 kilometres thick. It is estimated that over a period possibly less than a million years, 3–5 million cubic kilometres of lava erupted onto the surface of the Earth. This lava poured carbon dioxide into the atmosphere creating a super greenhouse effect. Temperatures rose rapidly and on land acid rain fell destroying plant life. The oceans heated up and underwent acidification as a result of absorbing carbon dioxide. As seen in Chapter 5, this forms carbonic acid in seawater, causing a change in ocean chemistry, which reduces the amount of dissolved calcium carbonate available for animals like coral to build their skeletons.

Oxygen was also severely reduced or lost altogether in the deep sea and waters poor in oxygen also upwelled into shallow waters. Furthermore, in the anoxic conditions microbial activity generated hydrogen sulphide, which also upwelled onto continental shelves poisoning marine life, a condition known as euxinia. Rising temperatures, ocean acidification and loss of oxygen are all symptoms of a severe disturbance to the carbon cycle. The chemical signatures of these hostile conditions are all present in rocks of marine origin that coincide with the extinction event. Marine life is very sensitive to changes in temperature as we have seen in recent decades with the occurrence of mass bleaching by reef-forming corals in response to ocean warming.

Ocean acidification can be a significant issue for organisms that build carbonate skeletons or shells as it reduces the amount of biologically available calcium carbonate in seawater. It also interferes with many other important functions in marine animals, even affecting their ability to 'smell' through detecting the chemical signatures of their environment, prey or predators. Hypoxia or anoxia simply kills organisms unable to obtain sufficient oxygen for respiration, the process by which we humans and much of the rest of multicellular life burns food to produce energy.

Following the Permian mass extinction, the marine fauna, as depicted by the fossil record, was characterized by a globally uniform low species diversity biota. The survivors struggled through several million years of hostile conditions before the Earth's climate finally began to stabilize and recovery took place. Some groups recovered faster than others. The ammonites increased diversity within the first two million years of the Triassic period. Coral reefs took a much longer time to reappear, and when they did they were formed by an entirely different group of corals, the scleractinia, or stony corals, which survive to today as the main reef builders. Other groups of animals, such as the trilobites, were less fortunate and were extinguished forever, although I must say it has always been my dream to find one of these animals in the deep sea: an undiscovered living fossil whose ancestors somehow made it through the Great Dying. I live in hope.

Deep-ocean anoxia (absence of oxygen) or hypoxia (reduced oxygen) has occurred throughout geological history and is usually associated with severe disturbances to the carbon cycle. The most recent event that caused a significant extinction in the deep sea was the Palaeocene-Eocene Thermal Maximum, some 55 million years ago. Up to half the species of shelled amoebas, the foraminifera, were lost in this event, and many other groups, including small crustaceans and even reef fish were affected. In the space of less than 20,000 years, global temperatures climbed 5°C–8°C causing ocean warming, acidification and anoxia. As yet it is uncertain what caused this event, but current evidence suggests again volcanic eruption and/or a massive release of methane into the atmosphere from marine clathrates. The latter are literally frozen methane, trapped under the seabed in conditions of low temperature and high pressure. Clathrate or methane hydrate appears as white ice, but samples retrieved from the seabed can be lit and burn. When conditions in the deep sea warm up clathrates can breakdown to release methane, which is a potent greenhouse gas in itself. One of my favourite marine animals is the ice worm *Hesiocaeca methanicola*, which is a bright-red, segmented bristly worm that actually lives on exposed methane hydrates on the continental slope in the Gulf of Mexico. The worms graze on bacteria living on the methane ice, which the bacteria themselves are consuming. To me the frightening thing about the PETM is that it is estimated that the rates of $CO_2$ release were about 200 million tonnes a year, possibly as high as a billion tonnes a year if the event was pulse-like as some studies suggest. Humans are currently releasing $CO_2$ at a rate of 10 billion tonnes a year, a geologically almost unprecedented rate and certainly not seen for hundreds of millions of years. The PETM is the closest event in geological history we have to the current global warming event we are driving.

I have already discussed the impacts of rising sea temperatures on the world's coral reefs, but is there evidence that the current disturbance in the Earth's carbon cycle is affecting levels of oxygen in the ocean? One of the most memorable animals from our expedition in the Arabian Sea was the squid. During the long evenings of the night watch between trawls or during coring operations, I would hang over the side of the ship looking at what the lights were attracting. The area around the gantry of the coring winch from where the box core was deployed was the best place as there were lots of lamps illuminating the working area of the deck creating a pool of light in the sea. Purple-backed squid, also known as flying squid, would move up from the depths and in from the dark edges of the light pool. I was absolutely transfixed by their movement. These large purple-red animals would shoot in, lancing their prey with a pair of long catching tentacles, enveloping it with the other thicker tentacles before diving into the darkness to consume it. Their ability to manoeuvre was extraordinary. They would suddenly stop, switch direction, swivel, dive down, jet upward, turn end over end, all propelled by a funnel through which they can explosively force water. One evening, as I was watching over the side of the vessel as usual, one of the squid jetted into the side of the ship momentarily stunning itself. Three or four dolphin fish swarmed in, even faster than the squid were able to swim, and ripped the unfortunate cephalopod to shreds in seconds. Now I knew why these animals only appeared during the night. The visual predators of the surface ocean were fast and deadly.

Recently headlines of newspapers in India and Pakistan have been reporting unusual numbers of purple-backed squid off their coasts in the Arabian Sea. There is mention that climate change is causing this sudden bounty of squid, although scientific studies from the 1990s suggest that the Arabian Sea does host dense populations of this particular species. Purple-backed squid are adapted to live in oxygen-depleted waters, and they hunt lantern fish at depths below 200 metres where conditions are hypoxic. Could a spread in oxygen-depleted waters be linked to the sudden perceived increase in purple-backs? At present, there is insufficient scientific evidence to answer this question. However, elsewhere in the world there are strong indications that oxygen-depleted water is spreading in area and, along with it, the distribution of one of the most spectacular and somewhat terrifying animals in the ocean, the jumbo squid. These animals grow up to 2 metres in length and weigh up to 50 kilograms. They are not the biggest squid in the world, but the larger ones, the colossal squid from the Southern Ocean (up to 14 metres long and 750 kilograms) and the giant squid from the deep sea (possibly more than 20 metres long), are rarely seen by humans. Jumbo squid are notorious for being extremely aggressive when hunting – they flash red and white when doing so, earning themselves the nickname 'red devils'. A casual glance at the internet will reveal reports and dramatized video of these animals ferociously attacking scuba divers, and there are even reports of them having killed fishermen off Mexico. While the veracity of these reports is difficult in most cases to confirm there is no doubt that encounters with these animals can be extremely intimidating, and scientists and film-makers trying to swim with them have taken to wearing chain mail, the same type as worn to defend against shark bites.

Squid, like all cephalopods (octopus, squid, nautilus and argo-nauts), have a beak made of a tough horny material called chitin, very similar to that which makes up the exoskeleton of insects, but much thicker. The bigger the squid, the larger the beak and the jumbo squid has a beak that in larger specimens could certainly cause significant injury to humans, even severing fingers and breaking bones. In addition, the tentacles of the jumbo squid are equipped with hundreds of suckers ringed with a row of small triangular teeth, also made of chitin. These are designed to dig into the flesh of their prey, gripping them firmly and propelling them towards the razor-sharp beak.

Like the purple-backed squid, which I saw for the first time off the coast of Oman, jumbo squid are adapted to live and hunt in oxygen-depleted zones, that in their case occur off the west coast of South America. However, following a major ocean warming event – the El Niño of 1997/1998 – the jumbo squid began to move northward and appear off areas such as central California.

El Niño is a natural phenomenon caused by the build-up of warm water in the western central Pacific. Periodically, the warm water flows back towards the east warming up the usually cold waters where the Humboldt Current prevails. The result of this is flooding in South America from torrential rains and mass die-offs of marine life acclimated to cold water. This region is usually extremely productive because of upwelling of nutrient-rich deep waters to the surface and it supports some of the largest fisheries in the world, mainly targeting anchovy. When El Niño occurs, everything, from the bottom of the food chain, including the anchovy, right up to predators, such as seabirds, is impacted. Warming from climate change is driving an increase in the frequency of El Niño events and in the proportion of these regarded as extreme.

The 1997/1998 event was certainly extreme and was the one that led to the loss of 16% of all the world's shallow coral reefs. It became apparent that the oxygen minimum zone off California and throughout the region was expanding. Hypoxic waters were extending closer to the ocean surface, causing the depth distribution of the prey of jumbo squid to become shallower. Studies using electronic tags showed that the jumbo squid spend most daylight hours at depths of around 250 metres, but at night they migrate towards the ocean surface. During this time the squid undertake dives into the oxygen-poor water at depths of around 300 metres, hunting animals such as the lantern fish and other squid, which themselves migrate towards the surface at night from the mesopelagic zone, the waters that lie between 200 and 1,000 metres, where some sunlight is detectable during the day. They also move into the shallower waters of the shelf, feeding on fish that dwell near the bottom, such as rockfish and hake. It would appear that the jumbo squid follow the shoaling of oxygen-poor waters as they have spread north. At the same time, other changes have been occurring in the ecosystem, including a decrease in the amount of hake in the region, possibly a result of environmental change, but also predation from the jumbo squid. Tuna, billfish and other surface ocean predators have also declined, potentially reducing predation pressure on the squid themselves. Among the factors that may have driven down the numbers of tuna and billfish is a reduction in the size of their available habitat. These vigorously swimming predators of the ocean require lots of oxygen. As waters poor in oxygen have spread towards the surface, the amount of richly oxygenated waters for them to swim in is being reduced, which means that they are literally being squeezed out from below. Put simply, a warming ocean causes a reduction in the mixing of oxygen-rich surface waters with deeper waters below. Warmer waters also contain less oxygen.

The squid are adapted to live in waters poor in oxygen so they have spread northward with the increase of warmer, low-oxygen waters. This is also probably driven by a shift in their prey, such as lantern fish, which take refuge in the oxygen-poor waters during the day. The expanding oxygen limited zone off California also means an expansion of habitat for other animals which permanently live in these waters including the vampire squid. This dark-red squid with haunting blue-grey eyes, a large pair of fins on its mantle, and skin stretching between its tentacles resembling an umbrella or hood, does look like something from mythical infernal regions. It is specially adapted to cope with life in oxygen-poor waters.

In the summer of 2002 surveys of the seabed off the Oregon coast revealed dead and dying fish and invertebrates on the seabed. Elsewhere in the region, moribund animals were washing up on the beaches. Fishermen were finding that three quarters of the crabs in their pots were dead. Scuba divers in depths of less than 25 metres of water found themselves besieged by unusually dense shoals of fish crowding into shallow waters. Oxygen-poor waters had flooded onto the continental shelf from the zone where deep water was upwelling just offshore. In 2006 an even more intense upwelling of oxygen-poor water occurred, again killing much of the marine life and allowing mats of bacteria that thrive in oxygen-poor water to grow across the seabed at depths of only 50 metres. Such events were not observed prior to the twenty-first century. This new phenomenon is a result of the shallowing and intensification of the oxygen minimum zone off the west coast of North America combined with local variation in ocean currents.

In 2002, nutrient-rich water entered the system from the north causing a burst of phytoplankton – marine algae – growth. This bloom was fed on by bacteria as it sank into the waters off Oregon, and as the bacteria metabolized this bounty of food they further reduced levels of oxygen. Seasonal hypoxia and the sporadic flooding of the coastal waters of Oregon with low-oxygen waters is predicted to increase in the future. This sinister new development has also had severe impacts for the oyster industry along the Pacific coast. As bacteria in the low-oxygen waters of the upwelling zone use up oxygen they produce $CO_2$, which, as I have mentioned before, is converted to carbonic acid when dissolved in seawater. This natural acidification occurs on top of acidification resulting from human $CO_2$ emissions and alters the chemistry of seawater causing it to become corrosive to calcium carbonate from which the shells and skeletons of many marine animals are formed. Oyster culture on the Pacific coast of North America depends on growing Pacific oysters for which the tiny juvenile animals, known as spat, are artificially raised in hatcheries. Oyster larvae need calcium carbonate to build a healthy shell. The upwelling events have caused a series of crashes in hatchery-produced larvae and a depression in the production of oyster spat. The result is a direct threat to a local industry worth nearly $300 million.

The western coast of North America is not the only place affected by declining oxygen. In the eastern tropical Atlantic an upwelling of nutrient-rich deep water occurs off the coast of West Africa. Here low-oxygen waters have been expanding towards the surface at the rate of up to 1 metre per year since 1960. As with waters off the coast of California and Oregon this has meant a squeeze of habitat in the upper ocean for animals that need good supplies of oxygen to survive. Marlin are 1 example. These fast-swimming predators are reputed to be able to reach speeds of 80 miles per hour, though these speeds are unrealistic and the reality is probably 20–30 miles per hour. This is still a very fast swimming speed and these animals, like many upper ocean predators, such as tuna, have a high oxygen demand. Electronic tags have demonstrated that these animals do not enter the oxygen-poor waters of the oxygen minimum zone. It is estimated that from 1960 to 2010 these animals have lost up to 15% of their habitat in the tropical north-east Atlantic as a result of expansion of the oxygen minimum layer.

Habitat compression has a variety of consequences for the ecosystems of the upper ocean. For predators like jumbo squid, which can tolerate low-oxygen waters but can also hunt in oxygenated surface waters, it might mean an increase in the density of prey. This may be one contributing factor to their spread in the eastern Pacific. For marlin, tuna and other fast-swimming species, it may make them more vulnerable to overfishing.

Climate change is not the only driver of oxygen depletion in our oceans. The green revolution that occurred between the 1930s and 1960s involved the development of new high-yielding varieties of crops, and the widespread increase of chemical fertilizers and pesticides, as already mentioned. The enormous increase in our capacity to produce food has been critical to supporting an expanding human population and it has been estimated that 40% of us are here today because of the use of chemical fertilizers. However, the use of nitrogen and phosphate fertilizers has been profligate and combined with poor land-management practices has led to a significant amount being washed off the land and ending up in streams and rivers, eventually finding its way to the coast. Along with nutrient-rich soil – also eroding from farmland – waste water, sewage and atmospheric inputs of nitrogen compounds from agriculture and burning of fossil fuels, the effect has been to fertilize coastal seas. This might, superficially, sound like a good thing. Fertilizers stimulate the growth of algae in the ocean and this means more food for the ecosystem. Unfortunately, however, the bloom of algae that results from such nutrient hyper-enrichment, also known as eutrophication, dies after a period of weeks and sinks below the surface where bacteria break it down, consuming the oxygen. This process mimics that of natural oxygen limited zones where the upwelling of nutrient-rich deep water stimulates algal growth instead. This may begin to manifest as the development of a sporadic occurrence of low oxygen levels, usually during the summer when waters are warm and the mixing of oxygen-rich surface waters with those below declines because of stratification.

These sporadic episodes of hypoxia are accompanied by the death of marine life in the zone, especially animals living on the seabed, but also fish and other swimming animals. As nutrient levels climb higher still, low oxygen levels become a seasonal event, occurring every year and creating what has been called an ocean dead zone. Areas where these dead zones have been described include the Baltic Sea, the Black Sea and the northern Gulf of Mexico. Here, the Mississippi River drains about 41% of the mainland mass of the United States and is a major contributor to an estimated average input of 1.6 million tonnes a year of nitrate flowing into the Gulf, most of which comes from agriculture. The result has been a seasonal dead zone which has grown in area from the 1970s to the present to about 20,000 square kilometres. Only the Baltic Sea dead zone is larger at 70,000 square kilometres. In the worst cases, hypoxia can become persistent, lasting all year.

Coastal dead zones have grown in number exponentially since the 1960s and now number over 400 globally. Because temperature has such an important influence on the oxygenation of the ocean, warming will make the coastal zone more vulnerable to the occurrence of hypoxia and the possibility of more dead zones developing. A few areas have shown some recovery where the use of chemical fertilizers has declined, so this is a problem we can solve through better management of agricultural practices. An example of this is the Black Sea where subsidies for chemical fertilizers vanished following the collapse of the Soviet Union. A hypoxic zone that had grown to more than 40,000 square kilometres in area by 1990 disappeared by 1995, although the ecosystem is still in recovery.

At times, the enrichment of coastal waters by fertilizers causes a bloom of algae that produce toxins. These are called harmful algal blooms (HABs). An estimated 200 or so species of algae produce toxic chemicals. Why they do so is often unknown: they may be involved in storing food or for protecting the algal cells from the harmful effects of exposure to ultraviolet radiation from the sun. In some cases, they may actually be produced to poison animals that feed on the algae. The blooms turn the sea in which they live red, hence the term 'red tides', or in some cases a lurid green. These tides can be poisonous to a variety of marine life including fish, crustaceans, molluscs and even animals as large as whales. Animals are either poisoned directly or through eating food contaminated with the algal toxins. Like certain organic pollutants, some of these algal toxins are accumulated up the food chain so predatory fish can concentrate the poisons in their flesh. Some of these toxins can even be fatal to humans who have ingested fish or shellfish contaminated by the harmful algae. Symptoms vary but often include vomiting and diarrhoea. There can also be neurological symptoms, which vary from tingling of the skin, to weakness and even changes in sensation so that cold things feel hot and hot things feel cold. In the worst cases, such as in amnesiac shellfish poisoning, caused by a substance called domoic acid, coma and death can result. And if you avoid this fate, poisoning victims can suffer permanent memory loss. Along the coasts of Florida, residents have reported respiratory problems and eye irritation during red tides. This suggests that wave action might actually cause aerosolization of algal toxins, harming those who breath them in. When I first heard of this I found it particularly disturbing. That human activity might be causing such a change in the coastal ocean that it might actually become hazardous to breathe in the sea air, is the stuff of nightmares. This is not to say that red tides never happened prior to modern times. A passage from the Old Testament (Exodus 7: 20–21) describes how Moses struck the waters of the Nile and the river turned to blood, killing the fish and smelling so foul the Egyptians could not drink from it. This sounds convincingly like a harmful algal bloom, although its occurrence in a river would be highly unusual. Historical accounts suggest that prior to invasion by Europeans, Native Americans kept watch for signs of red coloration or bioluminescence in the sea. Should these signs be seen, the local chiefs would forbid the eating of shellfish.

Nutrient enrichment as a result of human activity has increased the occurrence of HABs. Climate change, especially the effects of rising temperatures, may further increase the frequency with which coastal ecosystems are affected by these toxic blooms and may also cause them to spread geographically.

The fact that humans can affect levels of oxygenation in the ocean is something that most people have little awareness of. Yet, along with ocean warming and ocean acidification, it is one of the main symptoms of global climate change. However, it is also symptomatic of pollution of the ocean resulting from careless use of chemical fertilizers, intensive agriculture and poor land management as well as discharge of sewage and waste water into the ocean. The correlation of extinction events through history with low oxygen concentrations demonstrates that our oceans have always been on the edge of hypoxia or anoxia. Geological and biological processes appear to mediate the lurch between an ocean where oxygen-rich waters circulate from the poles to the deep sea, reappearing at upwelling zones and ventilating the ocean, and times when the world warms up, the oceans stagnate and life in the deep ocean suffocates.

I cannot help picturing our modern-day ocean circulation as the means by which the planet breathes. While a massive anoxic event may be a long way off, our rate of $CO_2$ emissions into the atmosphere appears from the fossil evidence to be unprecedented. Expansion of oxygen minimum zones is occurring globally, and a decline in the oxygen levels of the deep Southern Ocean has been detected – this is one of the main areas where oxygen-ated cold waters sink into the deep sea. This expansion, coupled with the alarming rise in coastal dead zones, is already having consequences for our oceans and for us. Fish stocks are damaged or have shifted as a result of changes in ocean oxygen levels and the shellfish we farm are being killed. Species that can take advantage of the spread of low-oxygen waters, like the jumbo squid, are invading waters where they were previously never seen. The diversity of marine coastal communities can be devastated by anoxic events. The seafood we gather from the sea can be contaminated by HABs, and perhaps even the sea air, previously considered healthy, can, on occasion, reportedly make you ill. The warnings from the fossil record are repeated and quite clear. Disturbances to the Earth's carbon cycle are bad news for the ocean. Warming, acidification and anoxia are three horsemen of the apocalypse that turn up again and again throughout geological time to reap their harvest of species, wiping out entire groups of animals. Perhaps a forth horseman is the occurrence of unexpected catastrophic events such as super-strength hurricanes and cyclones which we are seeing increasingly in the ocean today, but which are not so easily recorded in the fossil record. Recovery from these events can take millions of years, and when it happens the Earth is a very different place. Now it is we who are pumping billions of tonnes of $CO_2$ into the atmosphere, paving the way for another mass extinction event.

I find it remarkable when I am buttonholed at a meeting by someone who insists on telling me that the scientific evidence for human-induced climate change is somehow unfounded. They often claim that sea ice in the Antarctic is expanding or it was warmer than this sometime in the past. These people take very select pieces of information which, when isolated, can sound convincing to someone who is not appraised of the vast body of scientific knowledge that is all pointing towards human $CO_2$ emissions driving the current crisis. The example of the Antarctic is an interesting one. Yes, sea ice is on the increase overall, but this trend is driven mainly by cooling in the Ross Sea region of the Southern Ocean. Elsewhere in the Antarctic, some of the most rapid rises in temperature on Earth are being observed and sea ice duration has declined dramatically in recent decades.

Local variation in climate should not be taken as indicative of what is happening globally. Some of our politicians are simply dismissive of climate change, at least publicly. Donald Trump has on several occasions cited bouts of cold weather as evidence that global warming is not happening and has stated he will pull out of the Paris Climate Accord, although the terms of the agreement do not allow this until 2020. Of course, again, his policies in terms of maintaining or even increasing the use of fossil fuels are a nonsense when confronted with the reality of climate change.

I sincerely hope that by reading this book and seeing the damage being done to our oceans by warming, acidification and deoxygenation that you will support any efforts to reduce our dependence on hydrocarbons. Whether it be through switching off the lights when a room is not in use, cycling to work or to writing to your political representative, any action we take is significant. Climate change is a global emergency and the results of ignoring it will be nothing short of a calamity on the scale of a world war for humanity and for the environment. The warnings from our geological history are clear and mass extinction will be knocking on the door of the ocean very soon unless we can get carbon emissions under control quickly.

Dead zones may appear to be a more localized problem, but they are inextricably linked to food production and the use of the ocean as a free dump for our waste. Obviously we cannot support the seven billion and growing population of people on the planet without modern agriculture. However, we must use the knowledge we have gathered to better manage land and to use chemical fertilizers more sparingly than we have to date. Here technology has an important role to play in understanding where and when to apply fertilizers to prevent their waste and loss to the environment. Forest clearance and other forms of land use alteration have been responsible for driving an increase in run-off and soil erosion, all contributing to an increase in nutrients in the coastal zone. Ultimately these processes deplete the fertility of soils so improvement of water retention can only increase the sustainability and productivity of agricultural practices.

There is simply no excuse any more for dumping raw sewage and untreated waste water into the ocean. The ocean is not an endless hole where we can dispose of what we don't want for free. The perception that the ocean is vast and cannot be affected by our actions has been proven wrong time and again. Marine ecosystems are more delicately balanced than we could ever imagine. The balance of oxygen absorption by the ocean and its consumption by marine life, particularly microscopic organisms, is a great example of this. Changes in ocean mixing and rising temperatures are already causing changes that are rippling through ocean ecosystems. Add to this coastal zone eutrophication where these changes can be dramatic, albeit at smaller scales (still tens of thousands of square kilometres), then the myth of the invulnerability of the ocean is thoroughly dispelled. Industry and governments have to realize that they carry a responsibility to protect the natural environment and to maintain the services it performs for us.

The alternative? The oceans really will begin to run out of air.

# 8

# The Case for Hope:
## *Restoration and Recovery*

In the preceding chapters I have detailed some of the critical problems facing the ocean and what to do about them. What gives me most hope regarding the current plight of the ocean, though, is the miraculous ability of marine species to recover spectacularly from what has been the very brink of extinction. Populations of marine animals can be resilient with the result that compared to land, as far as we know, few species in the ocean have gone extinct. This means that much of the biodiversity of the ocean still survives and we have everything to play for! What is more, often all we have to do to encourage this phoenix-like behaviour is to protect marine ecosystems from damaging activities. Usually, this has meant reducing or banning fishing or hunting in specific places. Such marine reserves can be thought of as recovery zones, but they can also support the resilience of the surrounding ocean. This chapter is about my own experience of seeing such dramatic recoveries in marine ecosystems. It is one of the main reasons why, as a marine biologist, I can jump out of bed every morning with the enthusiasm to keep fighting for sustainable management of our ocean.

My first experience of this uplifting capacity of ocean life to bounce back was when visiting the island of South Georgia in the Atlantic sector of the Southern Ocean in the early part of 2009. The islands are a sight to behold, and the last time I visited we were on the RRS *James Clark Ross* on our way back from finding the hydrothermal vents on the East Scotia Ridge. The northern side of South Georgia presents dramatic mountains, snow-capped and rugged, ground apart by glaciers running down into the sea. There are also fjords and we gathered on the monkey island of the ship to watch as we progressed towards Grytviken, one of the old whaling stations on the island. Since we had already discovered the hydrothermal vents on the East Scotia Ridge, which meant the go-ahead for two further expeditions and lots of science exploration, we were all in good spirits and filled with excitement.

Paul Tyler, my friend and veteran deep-sea scientist from the University of Southampton, stood at the rails chattering about how he had always wanted to visit Shackleton's grave and see the old whaling station. The ship came to a halt, and a representative from the government of South Georgia and the South Sandwich Islands came aboard and explained to us the ground rules for visiting Grytviken: keep to the footpaths, don't disturb the wildlife and be careful to avoid being bitten by the seals. We were only partially listening – we were all a little too excited about seeing the island and, more to the point, about getting off the ship.

As we strode away from the low buildings of King Edward Point, which housed laboratories and offices, along a gravel path towards the old whaling station, we were in for a surprise. It seemed that every tussock of grass by the side of the track and all along the beach contained a young Antarctic fur seal. They would lunge out of the vegetation, mouths wide open, bearing needle-sharp teeth. It was an unnerving sight – I had never come across such vicious creatures. We gingerly approached the rusted boilers and chimneys of the whaling station and the seals were everywhere. They were a golden brown on the belly darkening to chestnut on the back with small ears, large black eyes and a muzzle dimpled with long whiskers. Like gangs of uncouth adolescents pumped with testosterone and looking for a fight at any opportunity, when they weren't lunging at us they lunged at each other, snarling the whole time. If we'd been inclined to take their aggressive reception personally, it was interesting to note that Georg Forster, the naturalist who, with his father, accompanied Cook on his discovery of South Georgia, had a similar experience of fur seals in 1775. He noted that the youngest pups not only barked and chased them, but also attempted to bite their legs. No outsiders were welcome, it would seem.

On the beach were two grey-painted whale-catching boats, rusting, but with harpoons silhouetted and pointing to the sky. *Petrel* was painted in black on the bows of the first one. We walked past huddles of king penguins, all looking very miserable as they were undergoing their first moult. This was the stage where the fluffy feathers of the chick were all replaced by the sleek water-proof feathers of the adult. Scruffy clumps of downy feathers were still stuck to them, but their adult feathers were almost all through, a beautiful black and yellow on the head, silvery grey, almost scaly on the back, and white on the front. Walking up to the graveyard of the old whaling station on the opposite side of the bay, we crossed deep-green tussock grass divided by deep-cut streams that were home to docile female elephant seals and their round and fat pups. All were covered in a foul smelling slurry, a mix of mud and their own faeces, and the skin of the females was peeling as they were undergoing an annual moult. Where the streams met the sea, huge pieces of yellowing broken whale bones lay around. Fractured jaw bones and broken ribs were clearly visible in the shallow, clear waters lapping the shore. It was the chilling evidence of a past massacre of animals, which many scientists, including myself, regard as sentient with a high level of intelligence.

We eventually entered the cemetery, which was marked by a low, whitewashed wooden fence framing the graves. Shackleton's grave could not be missed. A great crudely hewn slab of granite jutting out of the ground with the words of the poet Robert Browning in brass letters inscribed on the back: 'I hold that a man should strive to the uttermost for his life's set prize.'

Shackleton was the explorer who brought all his men back from the Antarctic after their ship, the *Endurance*, was crushed by ice in the Weddell Sea in November 1915. His men hauled three of the ship's boats, each weighing a tonne, across the ice to the open sea and then sailed to Elephant Island, a journey that in itself took five months. As the men hauled the boats across the ice floes they were completely exposed to everything the Antarctic could throw at them, from treacherous ice and temperatures in the minus twenties to ferocious wind-driven blizzards. For much of the time their feet were soaked and frostbite was a continual threat, as was the possibility of starvation. They were even attacked by leopard seals, fearsome predators of the pack ice that can grow up to 3.5 metres long, which prey on other seals and penguins. As the ice floes they were drifting on disintegrated the men jumped on the boats and made for land. They finally reached Elephant Island, which I have seen with my own eyes from the RRS *James Clark Ross* and which resembles a series of jagged, sharp black teeth sticking up from the ocean. Most of the men remained there while Shackleton and a hand-picked group of five sailors launched the *James Caird*, one of the 12-foot open boats, to cross the Scotia Sea to South Georgia. This was a treacherous journey, and the men faced the full savagery of hurricane-force winds in the Southern Ocean. Landing on the southern side of the island, three of them, led by Shackleton, climbed over the uncharted mountains of the interior equipped with 50 feet of rope and a carpenter's adze to make it down to the whaling station at Grytviken. Several months later, at the end of August 1916, Shackleton returned to rescue his men from Elephant Island. It was an incredible example of leadership and the ability of humans to endure the most horrible privation. Shackleton returned to South Georgia in 1922 on the *Quest*, a converted sealing vessel, but suffered a fatal heart attack the day after arriving at the island. At the request of his wife, Emily, he was buried in Grytviken, a place he has now become synonymous with. Shackleton is a hero to many Antarctic explorers and scientists alike. Paul, Rob Larter – the Chief Scientist on this cruise – and I stood around his grave. I broke out a silver-plated flask I had specifically brought on the trip for this purpose. I had filled it with Irish whiskey to celebrate that Shackleton was born in County Kildare in Ireland. Paul was smiling away and gave a short speech to pay tribute to the great man.

It was a fitting end to a special trip.

When we turned around to make our way back to the ship, the vision of the decaying factory stood like a haunted relic of some dreadful crime from the past, which of course it was. Thousands of whales were flensed and rendered down for their oil in South Georgia in the twentieth century to make soap and margarine. Grytviken was the first whaling base in South Georgia, but whales were not the first animals to be hunted on an industrial scale. Captain Cook discovered South Georgia in 1775, and in 1777 wrote an account of his second voyage in his diaries, later printed under the title *A Voyage Towards the South Pole, and Around the World,* including going ashore at Possession Bay in the northern part of the island: 'Seals, or sea bears, were pretty numerous. They were smaller than those at Staten Land; perhaps the most of those we saw were females, for the shores swarmed with young pups.'

Encouraged by Cook's words the first sealing ship, the *Lord Hawkesbury,* turned up in 1786, commanded by Thomas Delano. The target of the hunters were Antarctic fur seals. At the time there was a thriving market for seal pelts in Canton in

China where the silver fleeces were traded for spices, silk, the yellow fabric nankeen, chinaware and tea. Despite discussions in London as early as 1788 about the management of sealing operations in South Georgia, increasing numbers of vessels visited the islands in search of seals that crowded the breeding beaches. In 1800 Captain Edmund Fanning from Connecticut reached South Georgia on the 22-gun corvette *Aspasia* and the crew spent several months clubbing fur seals, leaving for Canton in February 1801 with 57,000 pelts. Seventeen other sealing gangs arrived later that year to reap another 65,000 pelts. By this time elephant seals were also being hunted for their blubber to render down to oil. The sealers camped on the islands and lived off other wildlife, taking birds' eggs, seals and other animals for food. Penguins were used mainly for fuel as their fatty bodies contained lots of oil.

In 1825 Captain James Weddell, an English explorer and later owner of a sealing vessel, estimated that by 1822 over a million fur seals had been taken from South Georgia and the species was nearly extinct. Fur seal populations made some recovery only to be decimated again in the 1870s when the sealers returned. Some illegal fur sealers still plundered South Georgia as late as 1907 but in 1909 fur seals became fully protected under the Seal Fisheries Ordinance, which was extended from the British administration in the Falkland Islands to the dependencies. It was too late, though. In 1915 a Norwegian sailor found a single male juvenile fur seal on a beach and killed it, apparently accidentally. More than 16 million seal pelts reached the market from around the world during this period, but of course this doesn't account for those on lost ships or for more-shady business transactions that were never recorded. It also doesn't account for the unborn seal pups who died when their mothers were slaughtered.

It is thought that the fur seals survived in refuges in the remote north-western parts of the island of South Georgia. In 1919 the captain of a whaling vessel reported seeing five fur seals in a place called Jordan Cove on Bird Island. The following year a report reached the magistrate of the islands of several seals on islands close to the entrance of King Haakon Bay in the south-west of South Georgia. In the 1920s there were sporadic reports of fur seals seen at sea around the islands, but most were discounted as being unreliable. It wasn't until 1933 that scientists, ironically on a sealing boat, encountered about 20 fur seals playing in the sea and sitting in tussock grass around a small bay on Bird Island. Subsequent reports during the 1930s were of tens of seals, then in 1956 Nigel Bonner, a seal scientist working for the Seal Research Unit in London, found fur seal colonies on Bird Island and the Willis Island group numbering between 8,000–12,000 animals excluding pups. Now there are over 4 million Antarctic fur seals on South Georgia, over 95% of the global population. The fur seals feed mainly on Antarctic krill, small shrimps that swarm in their billions along the Antarctic Peninsula, and out around the islands of South Georgia and the South Sandwich Islands. They also feed on fish and squid. A bountiful supply of food and full protection from exploitation for over a hundred years has allowed the population to come back from almost nothing. Walking on the beaches of Grytviken it is hard to believe that these feisty and fearless golden-furred seals were nearly exterminated from history. These days they are so large in number that they are hard to avoid, and so sharp of tooth that they have bitten several members of my research team and colleagues from other research institutions over the years. I can only admire the animal for its sheer unquenchable character and for coming back from the brink of extinction with such steel.

Fur seals are not the only animals that were ruthlessly hunted in the Southern Ocean. The whaling station at Grytviken was built in 1904 by the Compañía Argentina de Pesca, set up by four entrepreneurs, two Norwegian, one American and a Swede. Whaling had gone through a hard time because of the advent of electricity and competition from the petroleum industry for oil-based products. However, the invention of a process to convert whale oil to soap and margarine, alongside technological developments, especially the invention of the explosive harpoon, revived the industry. Within 10 years a full-scale whaling operation had developed in South Georgia. William Allardyce, the governor of the Falkland Islands and Dependencies, acted swiftly to try and stem the rapid growth of the industry, limiting the number of whale-catching vessels to two per factory ship and charging first a royalty per whale caught and then a flat licensing fee per factory ship. As the industry spread from a single whaling station to other locations around the island, further attempts were made by the governor at regulation through restricting the number of catcher vessels and then prohibiting the killing of calves or females with calves.

A growing tendency to overkill whales led to increasing

numbers of unprocessed or partially processed carcasses in the bays where whaling operations were based, so Allardyce legislated that all parts of carcasses should be used. In 1914 the number of whaling licences was frozen in an attempt to sustain the industry over a longer time period. The whaling industry resented such regulation, and so in 1925 the first factory ship with a stern slipway entered service in Antarctica. The advantage of this technology was that processing of whales could now easily take place at sea, freeing the whaling fleets from the waters of British territory and heralding a period of unprecedented slaughter. Within five years the number of whales killed rose from just over 14,000 to more than 40,000 per annum. New nations joined the British and Norwegians, including Japan, Germany and the Netherlands. Oversupply of whale products, coupled with the Great Depression, led to a crash in prices and economic catastrophe for the industry. Norway tried to establish a temporary prohibition on whaling in the Antarctic, but it was ignored by other states. Coordinated international efforts to further regulate whaling met with limited success, and in the 1937/1938 season over 46,000 whales were killed. The following season catches dropped notably because of overhunting, especially numbers of blue whales, the largest animal on Earth, and ceased altogether briefly during World War II.

The end of the war unfortunately saw the reestablishment of industrial whaling. The International Whaling Commission was established in 1949 and tried to establish catch limits on the basis of blue whale units, where one blue whale equals two fin whales, equals two and a half humpback whales, equals six sei whales.

However, what this meant was that whaling states simply targeted the largest and most valuable whales, successively depleting populations of blue whales, followed by fin whales and so on to the smaller species. There was also large-scale illegal and undocumented hunting going on by the Soviet whaling fleet. By the 1960s whale populations in the Antarctic were largely depleted – a situation exacerbated by the deployment of increasingly sophisticated technologies to locate them at sea, such as helicopters. It was only the pressure of non-governmental organizations like Greenpeace in the 1970s and 1980s that finally brought whaling almost to a halt. In 1986 a moratorium of whaling was brought in despite objections by Japan, which continued to hunt whales in the Southern Ocean for 'scientific purposes' – that is, until its recent decision to withdraw from the International Whaling Commission and resume commercial hunting in its own waters. In any case, over the course of the twentieth century, attempts to regulate whaling up to the point of the moratorium had largely failed. From 1904 to 1965 over 175,000 whales were killed and rendered in South Georgia, while between 1904 to 1978 more than 1.4 million were killed in Antarctica, and this is almost certainly an underestimate. The effects on the Southern Ocean whale populations were catastrophic, and species such as the blue whale were all but extirpated from the Antarctic.

Despite this, our journey on the RRS *James Cook* in January 2010 along the East Scotia Ridge was not only memorable for the incredible images of heaps of yeti crabs, snails and barnacles lying around the hydrothermal vents below – I had also simply never seen so many whales. The officers of the watch would call over the ship's intercom whenever they spotted these majestic creatures and members of the science party would rush outside to try to catch a glimpse of the leviathans. My first encounter with them was in the northern Scotia Sea where a large pod of fin whales swam at extremely high speed past the stern of the ship. Within minutes they were gone, heading south to feeding grounds no doubt. It is no wonder these animals are nicknamed the 'greyhound of the sea'.

A few days before my birthday we had our ROV, *Isis*, in the water. Maybe it was because of the noise of the vehicle or winch or because the ship was stationary, that two humpback whales turned up. We watched from the rails of the ship as the animals swam around, their great dark-grey backs with the stubby humped fin breaking the water's surface and sliding below. As we watched, the whales became more and more curious, coming right up to the ship and then holding their heads out of the water with their pectoral fins or flippers held out in front of them. The top of the head was a dark blue with tubercles scattered across while the lower part was speckled white and ribbed with folds where the whale can massively expand its mouth to engulf krill. Down the centre were clusters of encrusting white barnacles, yellowing on top and with a scattering of stalked barnacles among them. We were being watched. As one of the whales emerged from the water and flopped on its side, displaying a knobbly and barnacle-encrusted fin, I found myself looking into a large blue eye. It was a breath-catching moment. We were all delighted and to add to the spectacle, chinstrap penguins also joined, porpoising around the ship. One even found itself on top of *Isis* as it emerged, perhaps thinking the robot was a small yellow iceberg floating at the surface. Later that day I also spotted minke whales lunge feeding, jumping across the surface of the ocean to catch food. Never could I recollect seeing such a variety of whales in such numbers anywhere before.

Scientific studies suggest that whales are returning to the Antarctic in large numbers. In a survey to the north-west of the Antarctic Peninsula in 2013 it was estimated that there were nearly 4,000 humpback whales and over 5,000 fin whales. Skin biopsies of humpback whales from the Antarctic Peninsula region between 2010 and 2016 indicated average pregnancy levels of over 60% in mature females with one year reaching over 80%. This is a sure sign of a healthy and recovering population in a phase of rapid growth. There have even been sightings of blue whales in the waters of South Georgia again. This species is recovering more slowly as approximately 99% of the population was killed in the period of industrial whaling during the twentieth century. Understanding the population size and distribution of these animals is still very difficult, even though they are the largest animals on Earth. Hydrophones, listening devices that record the calls of whales, deployed in areas like the Weddell Sea, allow scientists to track where blue whales are. They are beginning to reveal some of the mysteries of where these elusive but gigantic animals spend their year.

While advising on the BBC's *Blue Planet II* I was asked to review the script of the final episode on human impacts on the ocean. The programme aimed to portray the grave situation related to the state of the ocean, but also to give examples of where interventions had led to recovery. One such example that particularly impressed me was the story of the Norwegian spring-spawning herring. The episode showed fishing vessels scooping up numerous herring alongside killer whales and humpback whales who were joining in the bonanza and gobbling up as many of these fish as they could. Herring are what are called a forage fish: these tend to be smaller fish that feed directly on plankton and which occur in huge abundance. They include species like anchovies and sardines, and they still support some of the largest fisheries in the world. The fish are eaten or are rendered down for fishmeal, which is used for animal feed in aquaculture and farming or for other purposes. Wherever vast forage fish populations are found – which are usually the highly productive parts of the ocean, like upwelling zones – they are a key species in the ecosystem. This is because they act as a conduit for energy and the flow of organic matter from the base of the food chain, algae and zooplankton, to aquatic predators, such as large fish, seals, seabirds and whales.

It is difficult to overstate the importance of herring, along with cod, in the history of northern Europe. In the early medieval times the spread of Christianity drove an increase in the consumption of fish. Religious abstinence from meat was required for up to 130 days of the year, and fish was a handy substitute. Originally the fish sought were mainly from freshwater and estuarine environments with wealthy people typically eating larger, more expensive fish such as pike, salmon and sturgeon. However, by the twelfth century, overfishing and changes to waterways led to a decline of sturgeon and other species that were popular food fish of the time. Fisherfolk turned to marine species, and as methods were developed to preserve fish by smoking or drying or by using salt or brine, it became possible to transport it inland to sell in towns and cities. Trading in fish was at the leading edge of the medieval commercial revolution, and herring were an important element of this wave of new enterprise. The Hanseatic League, a trading and defensive confederation of cities that grew on the southern edges of the Baltic, in what is now northern Germany, was partially founded on trading herring. The city of Lübeck, the capital of the League, sourced salt from Lüneberg, while the herring were imported from around the Baltic. Indeed, so important were herring, that in 1202 the Danes captured the entire merchant elite of Lübeck, as well as the city's fleet of ships, in Scania, where they were attending an annual autumn market to buy the fish.

In the late thirteenth century, the herring fishery in the southern Baltic collapsed. It is thought that a combination of environmental change, mainly linked to changes in coastal water quality and the clogging of coastal estuaries and bays with silt washing off farmland, combined with heavy fishing, brought to an end a stock that had been exploited for 400 years. This combination of environmental change and overfishing of stocks is one that has been repeated through to modern times and results from the position of herring and other forage fish in the food chain. In 1924, Alistair Hardy, an ex-head of the Department of Zoology in Oxford, mapped the diet of herring as they grew from larvae to adults. In the North Sea, larval herring feed on phytoplankton, the microscopic algae that are the primary producers

261

in the ocean, the very base of the food chain. As they grow into juveniles and then adults, the herring feed on zooplankton, the microscopic animals that graze on phytoplankton and which live suspended in waters with little power of movement. The most important food are tiny crustaceans, especially copepods, which resemble tiny water fleas. These basal layers of the food chain, the phytoplankton and zooplankton, are strongly influenced by the physical conditions of the ocean, such as temperature and nutrient concentrations.

When climate varies from year to year, or over longer times-cales – decades or even centuries, this can have a knock-on effect on plankton communities, reducing the abundance of algae or zooplankton or changing the species to ones less favourable as prey for the herring. Temperature will also affect the growth rates of the fish themselves. These bottom-up effects – environmental change leading to changes in plankton and then causing changes in fish populations – are the reason that herring and other forage fish species are prone to dramatic changes in population size. The mid-thirteenth century marked the shift from the Medieval Warm Period, which had lasted several hundred years, to the Little Ice Age, a period of unusually cool weather and advancing Atlantic sea ice that lasted until the nineteenth century. The switch from warm to cool conditions was marked by ferociously unstable weather in Europe, and as the Little Ice Age intensified, further collapses in herring stocks occurred, including those of the southern North Sea and off Scania, the southern tip of what is now Sweden. The loss of coastal stocks of herring led to increasing exploitation of herring stocks further offshore. The Dutch, who had developed particularly seaworthy vessels and advanced technology in herring processing and preservation, came to dominate the trade, eventually supplanting the Hanseatic League in the supply of the fish to northern Europe.

It would seem that collectively, humans are not very good at learning the lessons of the past. Norwegian spring-spawning herring was once one of the world's largest fish stocks. The fish feed far from the coast during the summer in the southern Barents Sea and central Norwegian Sea to the west of Norway. As winter comes, the fish migrate to fjords in northern Norway and then migrate south in spring to spawn on underwater banks off the coast of southern Norway. The eggs are released over rocky seabed at depths of 20–80 metres. The larval fish hatch and drift northward in the Norwegian Coastal Current, ending up in the Barents Sea where they grow and mature over a period of three to four years. These small, immature herring form an important food source for many predators in the Barents Sea including harp seals, puffins and common guillemots.

Norway's traditional fisheries for herring targeted the fish during the winter and also as they were undertaking their spring migration for spawning. The fishery initially used beach seine nets, but this switched to purse seine and drift nets from the mid-1920s onwards which were deployed in the open sea or at the entrance of fjords. This shift in fishing technology meant an increasingly intensive exploitation of the herring. A summer and autumn fishery for the adult herring also developed in the Norwegian Sea, largely to the north and north-east of Iceland, and mainly involving Norwegian and Icelandic fishing vessels.

Eventually, the Russians developed a summer fishery too. As these fisheries expanded, exploiting the herring throughout the year and at different periods of their annual migration between feeding and spawning grounds, the catching power of the fishing fleets also increased with developments in technology. It could not last.

By the 1950s the size of the spawning stock had reduced from 16 million tonnes to 7–10 million tonnes as a result of heavy fishing. Usually, such a decline would thin out fishing fleets as the declining economic viability drove out boats that were losing money. However, the Norwegian government, anxious to use fishing to support coastal communities, economically subsidized the fishing fleet during years when the catch was poor. As time progressed fishing technology was also continually improving and the size and power of fishing vessels increased.

In 1965, the International Council for the Exploration of the Sea (ICES), a scientific body that advises on sustainable exploitation of marine fish, reported on the Norwegian spring-spawning herring and advised that any reduction in fishing effort would reduce catches of herring. According to their findings, however, the impact of the fishery for juvenile herring, which was largely a reduction fishery where the fish are rendered down for oil or for animal feed, was unknown. By 1969, it was recognized that the fishery for young herring was a significant problem. Fishing that damages the replacement of new generations of young fish to the adult stock is known as recruitment overfishing. In this case, not only were adult stocks being severely reduced, but the young fish were also being caught before they could reach maturity and breed. Regarding the adult fish stock, the report concluded that, 'a further increase in fishing rate should probably be avoided and even some reduction of fishing be considered'. Among the laughable understatements in the annals of fishing history, this must rank near the top. In 1969 and 1970, catches collapsed. It turned out that the heavy overfishing of immature herring had led to almost no recruitment to the stock from 1966 onwards. The behaviour of the herring had also changed: they no longer migrated offshore to feed.

The summer and autumn fishery ended in 1969 and in 1971 the Norwegian winter fishery for herring was banned and there were drastic cuts to quotas for immature herring. Although very late in the day, the prohibition on the reduction fishery saved the remnants of the herring population spawned in 1969. Virtually no spawning herring were detected by fisheries surveys in 1970 and 1971. In the space of 30 years the stock had collapsed from its post-war high of 16 million tonnes to possibly as low as 10,000 tonnes. The economic effects of the collapse of the fishery was devastating for the fishing communities in Iceland, the Faroe Islands, Norway and Russia. The effect on herring predators must have been significant too, although they are not well documented.

By the mid-1970s slow recovery was evident with an estimated 100,000 tonnes of herring present off Norway. In 1983–1985 three relatively strong spawning events occurred for Norwegian spring-spawning herring, which by that time had reached a stock biomass of about 500,000 tonnes. This event had major consequences in the Barents Sea, which hosts one of the largest concentrations of seabirds in the world – approximately 20 million of them – many species of which nest along the northern

# The Deep

coasts of the Norwegian Sea and around the Barents Sea itself. Fish, including juvenile herring and another forage fish species called capelin, are a major component of the prey of many of the birds, seals and whales around this sea. Capelin are a silver fish, like herring, but the scales are much smaller and the overall appearance of the fish is more delicate. Capelin follow the front of melting sea ice as it retreats north feeding off swarms of zooplankton that in turn are feeding on the rich blooms of algae that occur in response to the sunlight from which they have been cut off during the winter. In the autumn, the capelin head back southward to spawn off northern Norway and Russia. Cod, many seabirds, seals and whales feed preferentially on capelin but will switch to juvenile herring or krill if the former are in short supply.

As a result of the strong recruitment of Norwegian spring-spawning herring in 1983 a large number of herring larvae entered the Barents Sea. Young herring feed on capelin larvae and so this abundance of herring larvae meant a drastic reduction of capelin. There was also substantial fishing pressure on the capelin populations at the time. Cod, which normally feed on capelin, switched to eating young herring, but it appears that there were insufficient numbers of these to replace the capelin. The cod began to starve and turned on their own kind, with the adult fish turning cannibalistic and eating the juveniles. The result was a major collapse in the cod stocks, again with significant economic impacts. This was bad, but the effects on other predators of capelin was catastrophic. Starving harp seals invaded the northern coast of Norway in search of food and died in their thousands by becoming entangled in the nets of fishermen. On Bear Island, between Norway and Svalbard, the common guillemot population collapsed in a single year from 1986 to 1987 from 245,000 breeding pairs to 36,000. The population has reportedly still not fully recovered. Other seabirds, such as puffins, were also affected. Again, this demonstrated that fluctuations in forage fish populations, compounded by fishing, could have serious consequences for marine ecosystems.

By the late 1980s to early 1990s the herring populations had recovered substantially, and normal migratory routes and patterns of spawning were re-established. In 1996, an agreement was reached between Norway, Russia, Iceland, the Faroe Islands and the European Union as to what levels the herring should be fished at and how to share out the allocation of catches or quotas.

Since then, spawning stocks of Norwegian spring-spawning herring have continued to rise and are now estimated to be in the region of 5 million tonnes —a slight decrease recently but still a relatively healthy population compared to the near extirpation of the stock in the region in the early 1970s.

The story of the Norwegian spring-spawning herring is an important one for us to learn from. It shows that if pressure is completely removed from a heavily fished species then it can recover, taking back its place in the ecosystem and providing food both for us and for marine predators. It also demonstrates that international cooperation is absolutely critical in establishing the conditions for a sustained recovery to take place of such a heavily overfished resource. However, even in a species that can reproduce as prolifically as herring, the recovery took 20 years.

Encouragingly, for deep-water species of fish such as the orange roughy, which grow slowly and have a low capacity to replace the population, recovery has still been seen in large stocks off New Zealand and Australia. It is yet to be seen whether these recoveries will be sustained because in the Australian case, recovery may be related to fish that hatched before heavy fishing depleted the stock.

While the story of the Norwegian spring-spawning herring is encouraging, recoveries such as these are not always the case. North-west Atlantic cod is another example of a fishery whose impact on humans can be traced far back through history. In 1496, John Cabot, a Venetian navigator, was commissioned by King Henry VII of England to explore the Atlantic to search for new lands. A year later, Cabot discovered Newfoundland, returning with stories that the cod were so numerous off the island that there was no need for nets; a basket lowered into the sea would return laden with fish. His son Sebastian Cabot claimed that fish were so plentiful that they sometimes halted the progress of his ship. With such stories of untold fishing riches, European vessels were soon making for the New World. It may seem incredible given the limited sailing technology of the time, but by 1501 the Portuguese were fishing in Newfoundland, followed by the Normans and Bretons by 1504. They were soon joined by the French and Basques, with English fishers joining in after 1565. Thus began a history of fishing that was to last for more than 500 years. Fishing for cod, haddock and other groundfish (fish that live close to the seabed) spread along the coast up to Labrador in the north and down through New England in the south. Fish would be caught, salted and dried in the summer and transported back to Europe at the end of the season. The fish were not only numerous but also huge in size, so much so that French vessels returning to Europe in the eighteenth century graded their fish in three weight brackets: great cod weighing 90–100 pounds, middling cod weighing 60–90 pounds and small cod weighing less than 60 pounds!

Over time, the cod fishery waxed and waned. Generally, there was an inshore fishery executed by small open boats and then the offshore fisheries on the Grand Banks, an area of shallow ground to the south-east of Newfoundland. At times over the history of the fishery it has shown signs of overexploitation with catches falling, sometimes correlating with unusual weather or sea conditions. However, a large industry existed through much of the 500 years, interrupted occasionally by wars between the colonial powers to which the fishing fleets belonged. The fishing industry was also closely tied to the growing of sugar cane by slaves in the Caribbean. Vessels would transport slaves to the New World and return to England with cargoes of salted fish. They would also carry trash fish from Newfoundland to the Caribbean to feed slaves, hence traditional Caribbean dishes that contain salt fish.

In the twentieth century, the advances in technology in terms of fishing and the preservation of fish leaped forward. After World War II, powerful pair trawlers and long lining vessels appeared, as well as factory or 'mother ships' that could receive, process and freeze fish catches at sea. Catches increased sharply through the 1950s and 1960s peaking at around 1.9 million tonnes of cod in

1968, the year in which I was born. This was almost four times the long-term average yearly catch between 1895 and 1945. There followed a serious decline in catches to 468,000 tonnes in 1976. On 1 January 1977 Canada declared a 200-mile exclusion zone where only Canadian vessels were allowed to fish in an attempt to keep out foreign fleets. This saw a small recovery in catches to about 700,000 tonnes in the early 1980s.

Despite the Department of Fisheries and Oceans of Canada throwing highly advanced methods of fisheries stock assessment at the problem, it seems that there was little realization that the fishing industry was eating into its own natural capital, the fish stock. The establishment of the exclusion zone and the over-optimistic estimates of the size of the cod stocks, as well as substantial subsidies from the government, encouraged the Canadian trawling fleet to continue to modernize to catch fish more efficiently. In addition to this, since Spain and Portugal had been prevented from deploying a large offshore fishing fleet in European waters, their fleets began to fish outside the Canadian exclusion zone but blatantly ignored the prescribed quotas. Catches were going unreported on both sides of the exclusion line and on a massive scale. Furthermore, there was strong evidence of a practice known as high-grading among the fleets. This is where small cod are discarded overboard in favour of larger fish because they fetch a higher price. At times in the 1980s so many fish were being brought into the fish processing plants that cod had to be dumped. The warning signs were coming in from all over. Inshore fishermen found that fish catches dropped and the seasonality of when the fish arrived changed dramatically, a winter and spring fishery becoming a summer one. Something was clearly wrong.

Despite enquiries by the Canadian government and various condemnations from scientific reports, the situation worsened. After 1989, catches plummeted, and in 1992 the fishery for cod was closed after almost no fish could be found on the Grand Banks. Newfoundland, and other areas of the coast dependent on cod fishing, faced an economic catastrophe. The Canadian government was forced to support communities entirely ruined economically by the collapse.

For many years there was no sign of recovery of the offshore cod stocks. Once these predators had been lost from the system, many of the species on which they preyed showed a dramatic increase in abundance, including lobsters, crabs and shrimp. These animals may have contributed to the lack of cod recovery, as under such conditions predator reversal can occur. The crabs and lobsters no longer kept in check by predation by adult cod fed on juvenile cod, preventing recovery of the stock. In addition, there is some evidence that a period of cooler temperatures caused significant changes in the ecosystem. This may have also contributed to the collapse of the cod populations, although the main cause was undoubtedly rampant overfishing. This was another example where the combination of environmental variability and overharvesting could bring a thriving fishery to its knees overnight. Now, there are finally signs that cod are returning after 25 years, possibly aided by a warming of the area and large numbers of capelin, a favourite prey of the cod. However, the pressure is constantly on to increase catches as soon as there is evidence of a strengthening of the stock. Whether fishers will continue to crop the cod populations and prevent their return remains to be seen.

Part of the issue with post-war fisheries management has been the rigorous application of mathematical models to manage fish populations through quota. As I have related, fish like herring and cod show fluctuations in population size from year to year depending on how many young fish are recruited to the adult population. This unpredictability has plagued fisheries for centuries. In 1957, Ray Beverton and Sidney Holt, working in a United Kingdom government fisheries research laboratory in Lowestoft on the North Sea coast, produced a seminal work on how to model exploited fish populations. One of the concepts within this model was that fish populations are self-limiting. When they get too large, they exceed the resources available to support their numbers and compete with each other, driving up mortality and reducing population size. Fishing could exploit this tendency of self-competition by skimming off a part of the population and actually increasing overall productivity of the stock by removing within-species competition. The point at which a stock became most productive for a fishery was known as the maximum sustainable yield.

This concept was economically highly appealing. A fishery could be treated as any other financial investment whereby investment (in fishing vessels) could predictably yield an annual revenue in terms of fish, but the stock would never be depleted. Beverton and Holt's equations, which were largely aimed at managing stocks of single species, came to dominate fisheries management for the next 30–40 years and are still influential today. However, in the words of Ray Beverton himself in a lecture regarding fisheries management written before his death, 'biology became subservient to maths, in both staffing and philosophy'.

The issues with this approach were many. First, it drove a very short-term view of fisheries aimed merely at predicting the quota for fishing fleets for the following year. Following the maximum sustainable yield philosophy, it was quite possible to fish down the portion of the population that made the largest contribution to the next generation of fish, particularly, in the case of large fish like cod, the old females that produce vast numbers of eggs. This recruitment overfishing contributed to many fishery collapses. Many of the parameters required to predict maximum sustainable yield were difficult to estimate and prone to error. In particular, estimation of the quantity of fish in a stock from the numbers caught in relation to fishing effort was difficult, as fishers often got better at catching fish over time as a result of technology advances and experience. Many aspects of the biology of fish, particularly the distribution of different populations of a species, were often ignored for expediency of data collection. Again, this was a major issue for north-west Atlantic cod stocks where offshore and inshore populations of fish existed that showed complex patterns of behaviour, such as migration. Environmental variation was still very difficult to account for in terms of impacts on recruitment. Wider ecosystem consequences of fishing and environmental change were beyond the interests or methods applied by highly quantitative fisheries biologists. This is still the case today. However, as a result of a series of catastrophic fisheries collapses over recent decades, biologists now model fish populations over much longer time periods with the aim of retaining a sufficiently large spawning stock to prevent recruitment failure even when environmental conditions reduce survival of young fish. It is worth noting, though, that even these approaches do not cater for the most difficult aspect of fisheries management, which is the prediction of human behaviour. Application of the most sophisticated fisheries models will be defeated by under-reporting of catches, sharp practice, such as high-grading, and illegal fishing.

As we look at the exploitation of marine animals over time, it is very clear that history has repeated itself over and over again since medieval times, through the Industrial Revolution to now. Once a resource is identified and a market developed for it, an industry rapidly develops and overexploitation occurs. Even where considerable investment has gone into attempting to sustainably manage resources, the drive to profit, from individual hunters and fishers to companies, has often undermined regulatory efforts. This has resulted in catastrophic collapses in populations of whales, seals, fish and shellfish. However, as we have seen, once exploitation has been prohibited, in many cases astonishing recovery of the previously depleted populations can take place. What if we could manage the ocean in a way that made it more resilient to exploitation of living resources, or other human pressures on it? One way of doing this is to simply set aside areas of the ocean as recovery or resilience zones, more commonly known as marine protected areas. My first experience as to how effective such a simple measure can be was on the Isle of Man while I was doing my PhD.

# The Case for Hope

On a clear day in early April 1990 I was preparing to dive off Port Erin Marine Laboratory's small research vessel, the *Sula*. We were below the cliffs that ran down to the Calf of Man, a small island sitting off the south-west tip of the Isle of Man. Our job was to undertake a video survey of the scallop exclusion zone, an area of the sea just west of Port Erin town where dredging for shellfish had been banned the year before. The fishermen had objected strongly to this idea, especially the locals operating off the jetty at Port Erin. Our relations with them were strained and rather unpredictable. They might enjoy a drink with the students in one of the local bars in town on the one hand, but on the other, more than once, they had driven their boats through our divers on the surface or over them while submerged. This caused much shouting and waving of fists at the nonchalant fishermen but had come close to flying fists when they were confronted on the dockside over their behaviour. They clearly viewed the sea as theirs and the exclusion zone was an affront to their right to fish where and when they pleased.

Scallops were suffering from the effects of being fished since the 1930s on the islands. Two species were targeted on the Isle of Man. The first was the king scallop, a large bivalve mollusc with a strongly ribbed and rather beautiful red-brown shell, streaked with paler colours, that was flat on one side and rounded on the other. They could grow as large as a dessert plate and sat in a small depression on the seabed, rounded-side down, so the flat half of the shell was flush with the surface of the sand or gravel in which they lived. The other was the queen scallop, known locally as queenies. These were smaller than the kings, with both halves of

275

the shell rounded, and a variety of colours from red to orange to pink, often patterned with streaks or radial lines. Queenies tend to just sit on the seabed rather than lie in depressions. Scallops were fished using dredges, a heavy iron frame with teeth that is dragged along the seabed, digging up the surface layers so that the scallops would be flushed from their pits and into the dredge bag, a pocket formed from coarse netting. Such a fishing method was highly damaging to the communities of animals living on the seabed. Queenies could be fished with lighter, non-toothed dredges or more conventional bottom trawls.

Mike, our ever-suffering diving officer, was in charge of the video camera. He was an experienced diver who worked hard diving with the PhD students and undergraduates to get their work done in all weathers and seasons whenever it was possible to get in the water. I was in charge of the compass to ensure that we swam our survey in a single direction and the surface marker buoy, a float on a reel of string that was towed behind the divers to let the boat know where we were at all times.

Mike was below me for most of the descent, dragged down by the weight of the camera. I followed, arms and legs out in a star shape as though skydiving. A blast of air in my suit and jacket brought me to a hover a metre above the seabed. We were deep, about 35 metres down, and the scene was lit with a soft yellow light. I took my compass bearings and we set off. Pushed along by a gentle current we crossed over fine gravel. Brittle stars tried to scuttle away from the alien visitors from above, some multi-coloured with banded arms, others pure black. There were bright scarlet cushion stars with yellowish lines marbling their surface.

I also saw a pelican's foot shell, a strangely shaped snail with a turret-like shell that flared out on one side looking rather like the webbed foot of a waterbird. And finally, there were king scallops. They were sitting in depressions in the gravel as expected, shells slightly gaping, revealing a fringe of fine tentacles, marbled white and brown soft tissue and dozens of eyes spaced evenly around the lip like tiny droplets of mercury.

It was fascinating to see how beautiful the animals looked in their native environment rather than clammed up in a basin awaiting the chef's knife. Life was not completely safe for them in the exclusion zone, though. Common sea stars crept across the seabed looking to trap the scallops in their pits and then digest the hapless molluscs by forcing their stomach into the shells. The odd large edible crab was also seen, residing in larger pits and no doubt also a threat to an unwary scallop. Remarkably both kings and queenies could swim a short distance away from these predators by flapping their valves together, creating jets of water that lifted them from the seabed and away from their attackers.

Over 10 years, by which time I had left the Isle of Man, these exclusion zone surveys continued. During that time, although there was an increase in the scallop population generally around the Isle of Man as the result of rising temperatures, the numbers of the animals increased much more rapidly inside the exclusion zone. The density of scallops at the legal size for fishing inside the exclusion zone rose to more than seven times that outside the protected area. The scallops were also on average older inside the exclusion zone and grew to a larger size than those outside in the fished areas. The seabed communities within the exclusion zone also changed, with more upright branched animals that live fixed to the seabed appearing. This probably benefited the very early stages of life for scallops where they metamorphose from a larva to a tiny version of the adult they will become, at which point they attach to branching, three-dimensionally complex animals or algae living on the seabed.

The benefits of this tiny spatial closure of the ocean – only two square kilometres – to the scallop populations, and therefore the fishery, was significant. Evidence suggested that the fishermen began to focus their dredging on the area around the exclusion zone, probably because of an overspill of adult scallops on to the surrounding seabed. Recruitment of young scallops to the fishing grounds to the west of the Isle of Man also increased in subsequent years. Large old scallops mean a much greater reproductive potential per area of seabed in the exclusion zone than outside. Observations suggested the eggs and larvae from the protected area were spreading out and helping to seed the fished areas.

While the fishing community on the Isle of Man was at first dubious of the efforts to close areas of the seabed to dredging, they were certainly convinced by the increase in catches. In 2008 Douglas Bay was closed to dredging to enhance recruitment of the Isle of Man scallop stocks even further. Other areas were closed to protect young scallops produced through aquaculture in an attempt to artificially seed the populations around the island. Most interestingly, the fishers themselves were given a portion of a specially managed area as part of a wider zone for marine conservation in Ramsey Bay on the north-eastern side of the island. The idea was to involve the fishing community directly in the management of the fishery. The hope was that if the fishermen themselves could see the benefits of the fisheries exclusion zones, then they would be much more likely to accept such measures and to actively participate in sustainable management of the fishery in a way that benefited the environment as well. They decided to selectively harvest an area of this zone where, based on surveys, a high concentration of scallops were found. Just two vessels were allowed to fish the area for a couple of days and the profits from fishing were shared among members of the Manx Fish Producers Organisation, a local fishing cooperative.

The Isle of Man marine reserves demonstrate how setting aside even a small area of the ocean can be of enormous benefit to commercially exploited species. In this case the reserves are indeed acting as marine recovery or resilience zones. Scallops are the type of animal that will benefit strongly even from a small reserve. This is because as adults they do not move very far and because the young scallops require complex seabed habitat to grow, which they can find in the reserve. Marine reserves, although often bigger than in the Isle of Man case, have also been shown to benefit fish populations time and time again, with the reason being that areas closed to fishing can allow stocks to regain a natural population structure. Like the scallops in the Isle of Man, fish are allowed to grow to a ripe old age and a large size and contribute an enormous number of eggs to annual spawning events. Fishing tends to remove these large, old animals from a population first, reducing the production of new fish for following generations and making populations more vulnerable to collapse under the combined effect of fishing and environmental variation. Marine reserves can be viewed as insurance for a fish stock, a mechanism for de-risking investment in a fishery.

The best reserves are those of a large size, which have been in place for a long time, are isolated, enforced well and where no fishing is allowed. One study has shown that areas of the coast that are fished have two thirds fewer fish than marine reserves with four or five of the features listed. On average such reserves have twice as many large fish species, five times the weight of large fish per unit area surveyed and 14 times the weight of sharks. Another study has indicated that the weight of fish within marine reserves, areas where no fishing is allowed, is on average 670% higher than in fished areas. In partially protected areas the weight of fish was 183% higher than in unprotected, fished areas.

Even for the most mobile of species, such as sharks, marine reserves can be beneficial. For example, the marine reserve around Palmyra Atoll in the Pacific has been shown, through tracking studies, to offer substantial protection to grey reef sharks. Two thirds of sharks tagged with satellite trackers were found to remain within the protected area, although some did move out of the reserve and about 2% were captured by fishing vessels. Studies from protected areas within the Great Barrier Reef have also shown that several species of shark spend a large proportion of their time within reserves. The effects of such reserves also trickle through the ecosystem. For example, in Fiji the effects of three small marine reserves (less than a square kilometre in area) was demonstrated as not only benefiting the species richness, numbers and size of fish but also in having 260%–280% higher coral cover than unprotected areas. On the Great Barrier Reef, coral populations within protected areas are substantially more stable, less vulnerable to disturbances, such as mass coral bleaching, storms or disease, and able to recover more rapidly from such disturbances should they occur.

Why is it that marine reserves, or strict no-take zones, are so beneficial to the entire marine ecosystem? The reasons are many and complex but two things are particularly important. The first is that the fish we usually hunt have important functions within ecosystems. Sharks and large fish, such as cod, are predators, and when they are removed the prey they suppress through their presence can dramatically increase in numbers. The response of crabs, lobsters and shrimp populations to the destruction of the north-west Atlantic cod stocks is an example of this. As noted, this ecosystem change may have actually significantly hampered the recovery of cod stocks once directed fishing was stopped. The presence of predators can even alter the behaviour of their prey. For example, in coral reef ecosystems off East Africa, the presence of groupers and other large predators drives sea urchins to hide in cracks and crevices in the reef during the day. In areas that are overfished, the predators disappear and urchin numbers not only increase, but they also brazenly graze the reef both at night and during the day, damaging the reef ecosystem. Removal of herbivorous fish can also be a problem on reef ecosystems. Such fish graze down algae on reefs, which are direct competitors for space with the reef-forming corals themselves. When overfishing of parrotfish and other herbivorous fish occurs it makes reef ecosystems more likely to switch to a less biologically diverse algae-dominated system. This is especially the case following a major disturbance event, such as mass coral bleaching. Once a reef has switched as a result of such a combination of overfishing and disturbance it can be extremely difficult for the ecosystem to flip back to its natural, coral-dominated state. Additional stressors, such as pollution from sewage or agricultural run-off, can further push the system towards the growth of algae, which benefit from the presence of high nitrate levels.

The other reason reserves can be so effective is that they preserve the structural integrity of the ecosystem. Many forms of fishing are highly destructive to the wider ecosystem, with the use of dredges for scallops or bottom trawls for fish like cod and orange roughy being prime examples. Destruction of the animals, algae or plants (seagrass) that live fixed to the seabed and which provide complex habitats for other organisms, has widespread impacts on the entire ecosystem. As we have seen, branching, treelike animals, the epifauna, living on the seabed, for example, provide important attachment points and shelter for tiny juvenile scallops, or spat. Dredging destroys the epifauna and reduces the survival rate of the young scallops, in turn reducing replacement of the following generation. In deep-sea ecosystems and coral reefs, the epifauna builds complex calcium carbonate structures over millennia, and these provide rich habitat for many other species that have important roles in the ecosystem. Destruction of such habitat is devastating, utterly changing the ecosystem, with severe impacts on biodiversity. Recovery of such systems is highly unlikely over timescales of relevance to humankind, and the result is often a low-complexity ecosystem with lower diversity and potentially less use to humankind. Reserves can maintain such critical habitats, as well as the biodiversity within them and important ecosystem functions, such as acting as spawning grounds or nurseries for other animals, or rich feeding grounds, including for fished species.

There are many other reasons for setting aside areas of the ocean from human exploitation, whether it is fishing or other forms of extraction. As we will see in the next chapter, marine ecosystems provide many services to humankind other than simply food. Ensuring that these other ecosystem services are preserved is essential in maintaining the Earth's life-support system, both for us and the rest of the biosphere. Marine reserves can offer great benefits in terms of resilience to natural catastrophes as well as direct and indirect human stresses imposed on the ocean. Take climate change, for example. Marine reserves may allow the biota to maintain healthy and resilient populations or to shift to keep track of favourable environmental conditions. Marine reserves can act as scientific reference areas through which change in the ocean resulting from what we do can be assessed in comparison to a baseline where no direct human exploitation is taking place. There is also a moral aspect to conservation. Future generations deserve the opportunity to see at least some of the ocean in as near as possible an intact and pristine state. Imagine if some of our most iconic marine ecosystems, coral reefs, kelp forests, deep-sea coral gardens, simply ceased to exist. This is a real prospect in many parts of the world and one that makes me shudder just to think about it.

# 9

# Natural Capital:

## *The Ocean's Endless Bounty*

It was 2011 and the International Programme of the State of the Ocean (IPSO), the small NGO I ran with Mirella von Linden-fells, had just published a report on ocean stresses and impacts. The report was written based on the conclusions of a small work-shop on the state of the ocean held in Somerville College in Oxford. Attending the workshop were leading marine scientists involved in marine ecology, climate change research, fisheries and pollution. It had been one of the most frightening meetings I have ever attended. Sitting in the comfort of a large, wooden-panelled and softly lit meeting room, scientists with different flavours of expertise described the latest advances in knowledge in their various areas with respect to changes in the ocean. At one point Jelle Bijma, an expert on climate research, described the effects of sea-level rise.

# The Deep

'Projections are suggesting that sea level might rise by more than a metre by the end of the century. Of course, this is the global average. Because of the distribution of the Earth's gravitational field there might be no change in somewhere like Britain, or even a slight decrease near Greenland, and a larger than one-metre rise if you are somewhere like Bangladesh.'

'Hang on a minute, Jelle,' I said 'Do you mean to say that where we all assume the average figures for sea-level rise are the same across the globe this is in fact not right?'

'Yes, that's what I mean.' Jelle nodded vigorously. 'The gravitational attraction of the huge mass of ice forming the Greenland ice cap draws the ocean towards it. As this begins to melt the gravitational attraction reduces, so the ocean moves away from Greenland and so sea level may actually drop in that area. In the UK it will be about the same but areas further south will suffer the most from sea-level rise.'

There were gasps around the table and hushed comments of, 'I never realized that.'

It was certainly a shock for me, and over the following two days there were several instances where the scientists around the table received information on the state of the ocean of which they were not aware. Scientists tend to specialize on their particular areas of expertise, so it was no surprise that an expert on coral reef ecology, for example, was not aware of the latest information on the state of global fish stocks, or in my case the latest predictions on sea-level rise. Both scientists and observers to the meeting, who were mainly from foundations and non-governmental organizations, ended the day's meeting in a state of intellectual

286

shock. We staggered out into the spring sunshine of the college gardens in the full knowledge that no matter how gloomy our individual perspectives of human impacts on the ocean had been before the meeting, the full picture was much more depressing.

The ocean was under severe stress from a variety of impacts: overfishing, destructive fishing practices, pollution, invasive species and climate change, with the latter responsible for warming, ocean acidification and deoxygenation. Some of these impacts combined so that the sum effects were greater than simply adding the cumulative effects from one stress to another. I called this effect negative synergy, where several impacts multiplied the levels of stress on marine ecosystems. Thinking about the future of the world and the implications for our children, the next generation, was enough to bring some of us close to tears.

The scientists who attended the workshop gave interviews to be used in the launch of the report, and several months later I found myself presenting the workshop's findings in front of a mixed audience of policymakers and experts at the UN headquarters in New York. Suddenly, the world was sitting up and listening to what we were saying and everyone wanted to hear more.

In New York, the phone didn't stop ringing. Sophie, a co-director of Communications INC, and I spent several mad days giving interviews to radio stations, newspaper reporters and other media. We would start at four o'clock in the morning doing Skype interviews for television and then answer calls between meetings while stuck in traffic in the familiar yellow taxis of the Big Apple. In an intergovernmental meeting in Europe, the UK delegate slammed the report onto the lectern and told the audience that it summed up why action on the ocean was urgent. There was a sense of ripples spreading across the world, a wake-up call for the oceans.

In a fascinating turn of events, the report also led me to Colombia for a meeting with the Kogi, one of the most ancient peoples of the world. It began with an email from a man called Alan Ereira immediately after the IPSO report was launched and was followed by a phone call as soon as I returned to the UK. Alan explained enthusiastically that he had made a film previously about the Tairona, the indigenous people of Colombia, who wanted to warn the Younger Brothers, meaning us and the entire outside world, that we were destroying the natural world on which they depended. They had decided that we had not listened when Alan had made a film the first time around and another film was needed to get the message across. Alan wanted to compare the beliefs of the Kogi, one of the tribes or groups of the Tairona, with respect to the natural world, with the modern scientific understanding of ecology. I was intrigued immediately. Alan had an excellent pedigree having directed, produced and written award-winning historical documentaries on the Battle of the Somme, the Armada and the Crusades. He had also worked alongside Terry Jones on *Medieval Lives* and *Barbarians*.

And so I found myself on a plane to Colombia and before I knew it I was sitting talking to Alan and his wife under the shade of trees in a hotel garden overlooked by the Sierra Nevada de Santa Marta mountains. Alan was a robustly built man with a full beard who brimmed with enthusiasm over his subject. We talked until dark about the Kogi who, from what I could gather, were virtually unique in the region in terms of retaining their indigenous culture almost completely. The Spanish conquistadors had arrived on the coast and founded the city of Santa Marta in the early sixteenth century and proceeded to subjugate the indigenous Tairona population with the intention of enslaving them and stealing their gold. The Spanish were equipped with armour and blades made of the famed Toledo steel as well as muskets, crossbows, horses and war dogs, which had allowed them to conquer other parts of Central and South America as well as areas of the Caribbean. The Tairona were no match for them militarily, however, and after decades of intermittent cooperation or conflict they realized that their entire way of life was under threat. The Spanish, who were devout Christians, decided that the Tairona were devil worshippers and morally utterly corrupt and that their culture was to be extinguished. The result, in 1599, was a revolt and a blockade of Santa Marta by the Tairona which was ruthlessly put down by reinforcements brought in by ship and followed by savage reprisals. The Kogi and others fled into the mountains and remained there to this day in almost complete isolation from the European invaders.

I first met the Kogi in a forest behind the sea on the coast. The area had obviously once been swamp but was drying up as the result of a huge plant to process coal having been constructed next to the sea. The Kogi stood underneath the trees dressed in plain white cloth trousers and tunics. It was immediately apparent that they were a very shy and retiring people and initial contact with the camera crew, the translator and various other strangers was a little tense and hesitant. The Kogis stood in a ragged line busily working a limey mixture in gourds called poporos, which each of them carried, scraping the paste on to the sides. The Kogi chew this paste with coca leaves constantly, a practice they believe connects them with nature. One of the Kogi Mamas – who act as the spiritual leaders of the Kogi – stepped forward to take charge of the delegation. His name was Mama Shibulata and he was Alan's main contact, and very much the star of the film. Mama Shibulata pointed to the coal plant and explained that when it was built the place they were standing in had begun to dry up. Now all the trees were dying because the conditions had changed.

The tribe led us through the undergrowth and out on to a sandy beach which was blackened with coal dust. Several workers were busy shovelling up the black sand. The company that owned the coal plant had heard we were coming to do some filming and had sent some men out to try to tidy up the black sand before it was recorded on camera. One of the Kogi picked up a snail shell to show me. The translator explained that this shell was what the Kogi gathered and burnt to make the calcareous paste that they held in their gourds. This was an important part of the Kogi culture. The shells were becoming more and more scarce along the coast, although the reasons for this were unclear.

A few days later I found myself following the Kogi, walking along a busy highway with huge trucks roaring past. We stepped down and through the edges of a mangrove forest, our loud footsteps sending small crabs scuttling away on the mud. When we reached our destination, we were faced with a scene I can only describe as resembling a World War I battlefield. What had once been a thriving mangrove forest and lagoon was now barely a few branchless dead and rotting trunks. I turned to Mama Shibulata who began to explain through words and gestures what had happened. He was saying that the land was like a woman's body, her head was in the mountains and her feet in the sea. When the road behind us had been built, it cut into the woman's body, which he indicated by drawing his hand across the side of his stomach. The connection was broken and the mangrove forest had begun to die. The blood of the woman had stopped flowing down her body to her feet. His description was an accurate analogy. The building of the causeway on which the road was placed had cut the supply of freshwater into the lagoonal system. Because of the warm climate, evaporation of the water in the lagoon made it increasingly salty until it killed the mangrove trees. The lagoon, known as the Ciénaga de Santa Marta, was a rich source of fish for local people and a haven for birds and other wildlife. Realizing the danger to this ecosystem, the government had made efforts to increase the flow of freshwater into the lagoon by cutting channels through the causeway on which the road was built. This had allowed some recovery of the wetlands to take place, although other problems were apparent, especially the dumping of rubbish along the shores.

One of the Kogi concepts I became fascinated with was that of association of certain areas with specific animals or plants. There was, for example, a Place of the Parrot, a Place of the Frog, a Place of the Coca Tree. The concept of the 'place' of an animal or plant was based on an understanding of population distribution that could only come from close observation. In ecology there is a concept that animals will occur at their highest densities where the richest resources upon which they depend, such as food, are found. Such areas must be important in the long-term survival of species as during periods of unfavourable conditions, for example, populations will go extinct in marginal habitat but remain in areas where resources endure. Such localities are the wellspring from which re-expansion of a population may occur after it has been affected by a natural catastrophe. The knowledge that the Tairona peoples had of the way that ecosystems worked and their sustainable management of them was truly impressive. They were capable of taking land that had been despoiled through logging or other forms of overexploitation and converting it back to healthy and productive forest using the simplest of tools. No wonder they were so concerned about what they saw the Younger Brother, us, doing in our ruthless extraction of resources and destruction of the natural environment. Yet they never showed anger, only a stoic humility and a willingness to try to impart their wisdom.

In the Kogi language, words translate to some concept of an object based on what it does, or how it relates to their way of life. One day, I asked what the word for the sea was. After some discussion and gesturing, I was told that the Kogi word for the sea translates to something like 'endless bounty' or 'endless food'. I haven't been able to get the concept out of my head since. Of course, to the Tairona people of hundreds of years ago the sea would have been an endless source of fish to eat and shells with which to charge their poporos. Thinking back, my own family had relied on the sea for their food and livelihoods for many generations. My grandfather had fished for lobster and crab for income and fish for food. My grandmother gleaned the rich shores of the west coast of Ireland for edible shellfish and seaweed to supplement the family's income. Even the vegetables were grown in soil that was fertilized with seaweed. Until recently the sea had provided a sustainable living for the fishing communities on the west coast of Ireland and in many other parts of the world.

Gleaning from the coast is an ancient way of gathering food that may even predate anatomically modern humans. There is evidence that *Homo erectus*, Java Man, thought to be an ancestor of modern humans, collected shellfish from the coast of South East Asia as long as 800,000 years ago. Neanderthals are thought to have foraged along the coastline of the Mediterranean between 125,000 and 30,000 years ago. Evidence also suggests that modern humans, *Homo sapiens*, in South Africa turned to foraging on the shore for molluscs 164,000 years ago. This move to eating marine protein coincided with population expansion and the development of tool use and may even have been linked to a spread of our ancestors along the coast of Africa and then beyond – moving into Asia between 70,000 to 50,000 years ago and then to islands in the south-west Pacific, including Papua New Guinea and Australia around 50,000 years ago. Human history is intimately connected with the ocean.

Even using the most primitive tools, our ancestors were capable of depleting local marine resources. The Santa Catalina Islands off the coast of modern-day California were colonized by humans about 13,000 years ago. By 8,000 years ago there is evidence that they had severely reduced the population of sea otters around the islands causing an increase in the prey of these animals, such as the red abalone, a type of marine snail. The shellfish in turn show a decrease in size in middens, ancient rubbish dumps, over a period of 10,000 years – another sign of overexploitation. As I described in the previous chapter humans were certainly capable of driving coastal populations of fish to collapse by the Middle Ages. It should be of no surprise, therefore, that the fleets of powerful fishing vessels that developed following World War II, equipped with modern nets or long lines of hooks, acoustic fish finders and ultra-accurate navigation equipment were capable of destroying the largest of fish stocks. In the rush to make profits over the last 50 years, governments and the fishing industry have often lost sight of the importance of the ocean in terms of nutrition. In a world where human population might reach 10 billion by 2050, fish are a critical component of food security. In 2013 fish supplied about 17% of the global population's consumption of animal protein. More than three billion people relied on fish in the same year to provide more than 20% of their animal protein uptake. In developing countries this figure can be higher than 50%. Fish are highly nutritious and even small quantities can provide essential amino acids, fats, such as omega-3 fatty acids, vitamins and minerals. Even in developed countries where diets are comparatively rich, fish can be an important source of trace elements, such as selenium. In 2014, around 80 million tonnes of fish were harvested from the sea. The ocean produces this food with no intervention from us humans, although there are obviously costs in going out and catching the fish or shellfish. But otherwise we don't have to put in any time or effort to produce this food. To put this into perspective, in 2016, just over 68 million tonnes of beef was produced by agriculture and about 117 million tonnes of poultry.

The ocean, however, is more than just a readily available supplier of food. It is an integral part of the Earth's life support system. An example is the global carbon cycle. The algae that make up the phytoplankton growing in the upper 200 metres of the ocean photosynthesize to convert carbon dioxide to the complex molecules that make up algal cells. Much of this carbon dioxide is recycled as algal cells are eaten by zooplankton and other organisms and are oxidized to produce energy, a small portion going to the growth of these grazing animals. However, about 1% of this organic material – in the form of dead cells, the bodies of dead zooplankton and other animals, faecal pellets and other detritus – sinks into the deep sea or is buried in the sediments of the seabed. When seen from a submersible this material looks like fluffy bits of snow drifting down in to the deep. Through this process, what I call the *passive biological pump*, $CO_2$ is drawn from the atmosphere and is locked away in the deep sea for hundreds to thousands of years. The ocean is the largest store of accessible carbon in the Earth system.

There is also an *active biological pump*, but we know much less about how important this is and how it works. The active biological pump is driven by the daily vertical migration of animals from the deep sea towards the surface, which I described in Chapter 4. The significance of this daily migration of billions of animals to and from the twilight zone to surface waters is that they actively transport carbon into the deep ocean. The animals feed at night in shallow water and then dive back into the darkness as day breaks to digest their food in the deep. It is a dangerous place, full of predators like the black devil anglerfish, *Melanocetus johnsonii*, I encountered this animal on my first oceanographic cruise on the RRS *Discovery* off the coast of Mauritania, West Africa, in 1993. The black devil anglerfish wafts a lure – which resembles a tempting, glowing morsel of food – around in the dark and attracts smaller predators that end up as dinner themselves. The anglerfish are a shiny, jet black and are literally all mouth and stomach, as once they encounter prey they must maximize their chances of trapping it and being able to swallow it! This is because food supply diminishes with depth. The jaws of the black devil angler are cavernous and packed with needle-sharp inwardly curving teeth. As the scientists gathered around, these fish would still be snapping at the spoons and forceps being used to lift out prized specimens from the bucket in which the cod end of the sampling net had been emptied. Together with its tiny eyes the black devil angler has a face that permanently appears to express rage. For someone who just loves unusual creatures it was a joy to see such a bizarre animal I had only before seen in books.

How much carbon is transported into the deep sea by active transport is poorly understood as until just the last few years we had very little idea of how much life resides there. Opening and closing nets, like the one that we used off Mauritania to catch the anglerfish, are traditionally used to study the animals living in the deep-water column. These tend to shred any gelatinous animals that are captured as their bodies are simply too delicate to survive the mechanical impact of the trawl gear and the abrasion of the nets. These gelatinous animals may make up a quarter of the mass of large animals in the twilight zone and yet they remain almost unstudied throughout large areas of the ocean because of this technical limitation. A few years ago a Norwegian scientist had the idea of observing a net being trawled through the mesopelagic using acoustics or sound. What he discovered was that the vast majority, possibly 90% of fish, swim out of the way of the net. These migrators include animals such as lantern fish or myctophids, small black or silver fish with elaborate patterns of bioluminescent organs called photophores, which they use for communication and camouflage. I have a picture in my mind of one of these trawls being dragged through the darkness glowing brightly like some sort of underwater funfair at night as it disturbs millions of deep-sea animals, which light up with bioluminescence. It probably leaves a great glowing trail behind it as well. It is not surprising most of the active animals simply swim out of the way. This rather simple observation indicated that scientists had underestimated the mass of fish in the twilight zone by a factor of at least 10. Instead of the original estimate of a billion tonnes of these fish in the ocean there may be more than 10 billion tonnes.

Both the passive and active biological pumps are superseded by the *solubility pump* in the ocean. Carbon dioxide dissolves in seawater forming carbonic acid, a purely physical process. The colder the water, the more $CO_2$ it absorbs, but other parameters, such as the relative amounts of $CO_2$ in the atmosphere and in the surface ocean, are also important. Part of the overturning circulation of the ocean, where cold, dense seawater sinks in to the deep ocean, a process known as convection, draws this dissolved $CO_2$ in to the deep sea, again sequestering it. Overall, the ocean absorbs about one third of the excess carbon dioxide that humans are emitting into the atmosphere through various activities. Without the ocean carbon pumps, it has been estimated that atmospheric $CO_2$ concentrations would be about 50% higher. The cost of this to the ocean is that the formation of carbonic acid is driving the process known as ocean acidification which is impacting coral reefs and other marine life, as described in Chapter 5.

But the value of our ocean goes beyond its ability to feed us and absorb $CO_2$. Its intrinsic climate control system is crucial to our survival. This works via thermohaline circulation, namely a transportation of the warm surface waters of the ocean towards the poles, which then cool, sink into the deep ocean and upwell again at lower latitudes. This massive and continuous movement of water and heat maintains global temperatures at the levels suitable for life on the planet – Earth's very own climate-control system. The ocean is also absorbing the excess heat generated in the atmosphere through the greenhouse effect, the trapping of the sun's heat in the atmosphere by gases like $CO_2$. Anybody who has a smart meter will realize how much energy it takes to heat even a small amount of water in a kettle. It is astonishing therefore that the upper ocean, from the surface down to 700 metres depth, has warmed almost everywhere measurements have been taken over the last 50 years. The northern hemisphere has shown the greatest warming, especially the North Atlantic. Although measurements are less plentiful from the deep ocean and go back less far in time, recent studies indicate temperatures are also rising from 700 metres to 2,000 metres depth. Overall, the ocean has absorbed an astonishing 93% of the excess heat trapped in the atmosphere by greenhouse gas emissions over the last 50 years. One group of scientists has estimated that if this excess heat was to be instantaneously transferred to the atmosphere it would lead to a temperature increase of 36°C. Under such conditions most of life on Earth could not survive.

The ocean can affect the atmosphere in the most surprising ways. There is a particular smell associated with the coast, a scent that characterizes 'sea air'. Much of this smell actually comes from a compound called dimethyl sulphide or DMS. This sulphurous compound is produced by seaweed that grows on the shore and in shallow waters and is also produced in large quantities by phytoplankton. It derives from dimethylsulfoniopropionate, which is a bit of a mouthful so scientists abbreviate this chemical name to DMSP. This substance is important to algae as it maintains the salt/water balance of cells, especially during times of stress. Some of it leaks from the algae, but much is released when the algae die naturally or when they are attacked by viruses or fed upon by grazing animals. When the DMSP is released, it is broken down by bacteria to form DMS, which is released to the atmosphere from the surface of the ocean. Once in the atmosphere DMS is oxidized to sulphate and sulfonate and this is where the miraculous connection between marine algae and our weather is made. These compounds can act as cloud-condensation nuclei, causing water vapour to condense into droplets around them and form clouds. Clouds reflect sunlight and thus can reduce the warming effect on the atmosphere. The thought that the tiny single cells of phytoplankton drifting in the ocean can control cloud formation and alter the albedo – or reflectance – of the atmosphere I find, intellectually, deeply satisfying. It is living evidence of the Gaia concept, the idea that the Earth as a whole, including its physical and biological components, functions as though it is a single entity, self-regulating to maintain comfortable conditions for all life, including ourselves.

Coincidentally it was the scientist who first conceived of the Gaia hypothesis, Professor James Lovelock, who went in search of a missing component of the Earth's sulphur cycle and demonstrated the importance of DMS production in the ocean. Without the release of sulphur, which is a critical element for life, into the atmosphere, it would have washed off the land into the sea and become lost to the ocean. It is by the biological production of DMS that the sulphur cycle is closed and the element returned to terrestrial ecosystems. Lovelock was also a co-author of the scientific paper in 1987 that drew together the evidence related to phytoplankton, DMS production and cloud-condensation nuclei. Recently it has been demonstrated that when coral reefs are under thermal stress during hot, calm weather, they also produce large quantities of DMS. This is produced by seaweed growing on the reefs and also the symbiotic algal cells, the zooxanthellae, in the reef-forming corals themselves as discussed in Chapter 5. Satellite observations have indicated that the DMS leads to cloud formation over the reefs, helping to reduce the amount of the sun's radiation reaching the ocean surface. I find the whole idea that reef-forming corals can control the weather to protect themselves through producing this gas simply wonderful, and so elegant.

So, with all of this in mind, what if we could put a monetary value on the services the ocean provides? Valuing 'natural capital', as it has been termed, has become a popular way of trying to get politicians to think about what the ocean does for us so that they are encouraged to account for it when making decisions. Coral reefs, for example, cover less than 0.1% of the area of the ocean, but they are considered by some to be the most valuable natural ecosystem on Earth per unit area. In 2014, it was estimated that a single hectare of coral reef is worth more than $350,000 per year and collectively they represent an annual monetary value to humankind of over $10 trillion. This is greater than the estimated value of tropical or temperate forests. A report from Deloitte Australia in 2017 estimated that the value of the Great Barrier Reef as an asset was in the region of $56 billion, bringing $6.4 billion to the Australian economy every year and employing more than 60,000 people. The report only considered tourism, fishing, recreation and scientific research. It by no means considered all ecosystem services performed by coral reefs.

In 2004 and 2005 the Indian Ocean tsunami followed by Hurricane Katrina demonstrated the importance of coastal eco-systems in the protection of coastal infrastructure and people from extreme flooding events. For example, reefs protect the coastline from storms and hurricanes reducing wave height by 70% and protecting billions of dollars in coastal infrastructure globally, not to mention people's homes and families. Overall, protection of the land, or prevention of its erosion, may be the most valuable ecosystem service provided per unit area of coral reef. Other coastal ecosystems, such as salt marshes, mangrove forests, seagrass beds and algal forests, perform a similar service providing greater or lesser levels of protection. These ecosystems are becoming more important as coastal human populations grow and climate change leads to a higher frequency of more extreme events. In 2017 alone, the USA suffered 16 weather and climate-related disasters, including hurricanes and severe storms, each of which caused more than $1 billion of damage, adding up to a total cost of more than $300 billion.

The ocean also harbours resources we do not yet fully understand, with genetic resources a case in point. Marine organisms, such as algae, sponges, corals and a wide range of other invertebrates, produce a range of chemicals or secondary metabolites. The organisms use these for a variety of purposes, for example as sunscreens, defensive toxins to prevent themselves from being eaten, or as antibiotics to prevent infection by pathogenic bacteria or viruses. Even in ancient times, consumption of seaweed in Japan and China was recommended for the prevention of the occurrence of goitre, a disease associated with a lack of iodine which causes swelling of the thyroid gland in the neck. We now know, of course, that seaweed is rich in iodine. At the start of the twentieth century it was known that roasted marine sponge contained iodine and could be used to treat the same disease. Cod liver oil has been used as a dietary supplement for more than a century as a result of its richness in vitamins A and D and omega-3 oils. Vitamin D deficiency is a cause of rickets, a disease that causes soft bones, deformed limbs and other developmental issues, which was rife in poor communities in Britain during the Victorian era.

In the 1950s active drug ingredients were identified in a Caribbean marine sponge, *Tectitethya crypta*. This species was discovered in 1945 in Florida by two scientists, Werner Bergman and Max de Laubenfels. Bergman was studying steroid-like compounds in marine organisms. When they boiled the sponge up they found a large amount of a crystalline substance formed. The substance showed similar physical properties to thymidine, a component of DNA, the molecule that makes up the genetic code of life. Bergman named it spongothymidine and later found another substance, which he called spongouridine, with similar properties to uridine, a component of RNA, another information-carrying molecule found in all life including us. These substances occurred in large amounts in a free state in the sponges and were found to inhibit the replication of DNA and RNA. Known as nucleosides, they were being produced by the sponge as a form of chemical defence. An artificial form of the chemical, arabinosyl cytosine, was synthesized and found to cure cancer cells in rats. This substance was licensed as a drug for cancer treatment, cytarabine, in 1969. The substance, named ara-C for short, is now used as a cell replication inhibitor in the treatment of several leukaemias and non-Hodgkin's lymphoma.

The hunt for marine natural products is now pursued more vigorously and systematically as countries around the world have realized that their coastal seas may harbour riches in the form of new drugs, cosmetic ingredients (cosmeceuticals) or food supplements (nutraceuticals). More than 20,000 marine natural products with unique structures have been identified, with the richest sources being sponges, microbes and cnidarians (corals, anemones, jellyfish, sea firs). The route to approval as drugs for substances which show therapeutic properties is long and expensive, involving four levels of trials with additional monitoring after a drug is approved for use. As a result, of the thousands of compounds being investigated, only a handful have been approved for use as medicines today. These include a range of drugs derived from sponges, sea squirts and a sea slug for the treatment of cancer, and a treatment for neuropathic pain isolated from the venom of a cone shell.

Because of the extremely high costs of trials and the lengthy process of gaining government approval for use of marine natural products as therapeutic drugs, many companies have looked to other uses for marine compounds. Cosmeceuticals are active ingredients used in cosmetics products. Compounds have been derived from unusual sources, including bacteria from deep-sea hydrothermal vents in the Atlantic and Pacific, and from intertidal environments in the Antarctic. The bacteria are grown through fermentation and the extracts derived from them are reported to soothe or prevent irritation of the skin when exposed to chemicals, friction or UV radiation, to act as skin restoratives and to hydrate the skin, preventing wrinkles or lines. Corals, such as the Caribbean sea whip, have been found to produce a suite of compounds called pseudopterosins which have been used to prevent skin irritation in some cosmetics. Microalgae and seaweeds are also a source of compounds that are reputed to confer various benefits, such as anti-ageing, anti-inflammatory or restorative properties for the skin. We also eat many marine derived products as health supplements, such as omega-3 oils in the form of, for example, cod liver oil. There are also food additives such as carrageenans, which are a group of sulphated sugars derived from red seaweeds used commonly as gelling agents or stabilizers in foods. The name is derived from caragheen, the Gaelic word for the seaweed commonly known as Irish moss, which my grandmother boiled up with milk to make a blancmange.

As well as use in the pharmaceutical, cosmetics and food industries, marine organisms have been the source of important chemicals used in research and industrial biotechnology. One example is green fluorescent protein. This protein naturally occurs in the jellyfish *Aequorea victoria* and produces a ring of green bioluminescence around the edge of the umbrella of the animal. Osamu Shimomura was a Japanese scientist who survived the atomic bomb dropped on Nagasaki as a young boy and worked at Princeton University in the USA in the 1960s. He and his colleagues collected over 10,000 of the unfortunate jellies in Puget Sound and cut off the ring of tissue from around the edge of the umbrella to collect the proteins responsible for the bioluminescence. The bioluminescent protein was called aequorin and glowed blue, but there was a second protein that fluoresced green under blue or UV light. This was the substance that became known as green fluorescent protein or GFP. Thirty years and 850,000 jellyfish later, the DNA sequence for GFP was discovered. The protein was of quite a small size and was the product of a single gene. This meant it could be artificially inserted next to other genes in an organism and wherever the adjacent gene was expressed, the GFP gene was also expressed. In other words, tagging a gene in an animal with GFP allowed scientists to determine where the gene was active and when it was switched on and off simply by whether the organism or a structure within it glowed green when exposed to blue or ultraviolet light. This was an enormous scientific advance and the uses of GFP in tracking the activity of genes or in tracking labelled cells have been many and varied. GFPs have been used, for example, to trace the circuitry of the brain, the entry of viruses or bacteria into tissues and cells and the regeneration of organs, such as the kidney. They have also helped to track the processes involved in growth and development of organisms. Subsequently fluorescent proteins of different colours were developed by mutating the original GFP gene or by looking for equivalents in other animals. Red fluorescent protein was derived from a coral for example. Shimomura and his colleagues won the Nobel Prize for Chemistry in 2008 as a result of their pioneering work.

Biotechnology, artificial intelligence and web-based applications are all viewed as features of the fourth industrial revolution. Genes can now be discovered *in-silico*, in other words by dredging databases of DNA sequences from marine organisms to identify the coding regions. In this new, rapidly moving age of the internet-of-things, however, the hunt for new biotechnologies, whether they are for drugs, cosmetics, food stuffs or other industrial uses, has taken on a concerning aspect. A study in 2018 identified that 12,998 genetic sequences from 862 marine species were associated with patents. These sequences were derived mainly from microbes, but also come from groups such as sea squirts and iconic animals, such as sperm whales and manta rays. Eighty-four per cent of all patents were registered by private companies, with a further 12% by private and publicly funded universities and government bodies. Non-profit research institutions, hospitals and individuals accounted for the rest. A single corporation, BASF, a German-based transnational and the world's largest chemical manufacturer, had registered 47% of all these patents. Three countries, Germany, the USA and Japan registered nearly three quarters of all patents while 165 countries have not registered any patents at all. Such an appropriation of the genetic wealth of the ocean is startling and raises many questions regarding equity among nations when it comes to benefiting from this new area of wealth generation, not to mention the oligopoly among the major chemical manufacturing companies in the world.

Understandably, developing countries, many of which lie in the biodiversity-rich regions of the world, have become alarmed about the possibility of bioprospecting on land and in the sea by external actors, leading to the loss of opportunities to benefit from such genetic wealth themselves. In 2010, a supplementary agreement to the Convention on Biological Diversity, known as the Nagoya Protocol, was established to try to regulate access to any state's biological wealth. It establishes a framework to ensure that a state working with industry, universities or other entities, receives a fair and equitable share of any benefits arising from genetic resources. It does not, however, extend to the high seas, the areas beyond national jurisdiction, usually more than 200 nautical miles from the coast. In 2017 a decision was made to initiate negotiations around a new implementing agreement for the United Nations Convention on the Law of the Sea (UNCLOS) to deal specifically with the protection of biodiversity beyond national jurisdiction (BBNJ). Given that this region effectively belongs to all of us, the intention is to include a package within these new regulations that deals with equitable sharing of the benefits of marine genetic resources outside of the maritime zones of coastal states. How this will apply in practice is unclear at present. However, the development of new industries has raised its head in another area where the ocean has potential to provide us with resources.

Our appetite for new technologies is driving the demand for some of the rarer minerals on Earth. Mobile phones, video monitors and renewable technologies, such as photovoltaic cells, wind turbines and batteries all require elements that are relatively rare on or near to the Earth's surface. These include such exotic sounding metals as bismuth, europium, neodymium, niobium, palladium, tantalum, tellurium, terbium and yttrium and some more familiar elements, such as cobalt, lithium, platinum and tungsten. In 2010, about 60 kilograms of tantalum, 510 kilograms of platinum, 22.5 tonnes of palladium, 51 tonnes of gold, 525 tonnes of silver and 24,000 tonnes of copper were used to manufacture about 1.5 billion mobile phones. This need for metals, including the rarer elements required for modern technology, is now driving states to look for minerals in the deep ocean. Already diamonds are excavated from the ocean floor off the coast of Namibia at a depth of 150 metres. Six vessels operated by the De Beers group of companies in partnership with the government of Namibia are involved in the mining programme. In 2015 more than a million carats of diamonds were dredged from the seabed and the operation is generating more than $10 billion Namibian dollars in revenue annually.

In 2017 Japan announced that it had undertaken the world's first commercial-scale deep-sea mining operation at a depth of 1,600 metres off the coast of Okinawa, recovering enough zinc to supply the country for a year as well as quantities of gold, copper and lead. There are several different types of mineral deposit of interest to mining companies in the ocean. Seabed massive sulphides are associated with hydrothermal vents. These are the accumulations of metal sulphides that precipitate from the vent fluids when they mix with cold seawater – the process that forms the iconic 'black smokers' thousands of metres below the ocean surface. Vents are usually associated with mid-ocean ridges, volcanic island arcs and hotspots that generate chains or clusters of seamounts and islands on oceanic tectonic plates, such as the Hawaiian Islands.

There are also manganese nodules or polymetallic nodules that form on the abyssal plains over millions of years. The nodules are mineral concretions, 1–12 centimetres across, that lie partially buried on the seafloor at depths between 3,000 and 6,000 metres. The nodules generally develop from some form of organic material, but are formed mainly of fine particles of iron and manganese. The iron has a slight positive charge and attracts elements such as molybdenum, vanadium and arsenic, as well as rare earth elements, while the manganese particles are negatively charged and attract metals such as cobalt, nickel and copper.

There are also cobalt-rich ferro-manganese crusts that form mineral pavements on seamounts, ridges and underwater plateaux from depths of 400 to 7,000 metres. These pavements are created by precipitation of minerals onto bare rock and other sediment-free surfaces and range from 1 millimetre to 26 centimetres in thickness. The richest zone for development of these crusts appears to be between 800 and 2,500 metres depth. These deposits contain high concentrations of iron and manganese, but also cobalt, platinum, tellurium and rare earth elements.

There are also metal-rich sediments, such as those which occur in deep parts of the Red Sea. Here, extremely salty water formed by the dissolution of salt-rich rocks sinks to the bottom of the ocean, forming brine layers over the seabed. Sulphide-rich hydrothermal fluids mix with the brine, and the sulphides react with metals in the seawater and sink to the seabed to form a metal-rich mud. The brine is largely anoxic and very hostile to life. There are also phosphorite deposits which are a potential source of phosphate for agricultural fertilizer. These form in areas of high surface primary production where waters become hypoxic or anoxic because of bacterial activity (see Chapter 7). Under such conditions bacteria can accumulate phosphate that they can release in such high concentrations that it leads to the formation of calcium phosphate on the seabed. Dying fish and other chemical reactions can add to the formation of phosphorites and over time there must also be an alternation between a seabed covered in anoxic mud and then periods of winnowing or erosion of the seabed. The phosphorites form as pellets, nodules, cobbles, crusts, slabs or cemented sediment. Finally, there are also iron ore sands which contain, of course, iron, but also titanium, garnets and gold. Planning is in an advanced stage for mining such sand off the coast of New Zealand.

All these deep-sea mineral deposits are associated with different ecosystems, and mining them may affect different habitats and species with different distributions and ecology. Deep-sea hydrothermal vents form island-like habitats for animals that are specially adapted to the conditions found at these locations. More than 70% of these animals are found nowhere else; in other words they are endemic to vents. The vent ecosystems are generally very small in comparison to the surrounding ocean, and connectivity of populations of animals to those on other hydrothermal vents depends on a number of factors. Geology is important as the rate of formation of new seabed or oceanic plate at a mid-ocean ridge, known as the spreading rate, determines the density of vents along a ridge. Fast-spreading ridges can have a vent site every 5 kilometres whereas on slow-spreading ridges the density of vents is more like every 100–350 kilometres. However, something like 86% of the potentially economically viable seabed massive sulphide deposits by virtue of their size are on slow-spreading ridges because vents in such settings last for much longer than on fast-spreading ridges, perhaps hundreds or even thousands of years compared to tens of years. This means the mineral deposits are much larger than on fast-spreading ridges.

Seamounts are also island-like habitats, and although levels of endemism are not high as in hydrothermal vents they are species-rich and diverse habitats. They are also important feeding grounds for many oceanic predators and are used as breeding sites or navigational waypoints by some animals. Large seamounts cover in the region of 4.7% of the seafloor, so these are also important habitats in the deep sea.

The abyssal plains, where manganese nodules are found, are vast areas of the deep sea covered in fine sediment where levels of natural disturbance are very low. Experiments have shown that the seabed fauna in these areas does not show full recovery from small-scale disturbance after 26 years. The epifauna, the animals living on the seabed, often attached to the nodules, are the component of the abyssal fauna that appear least capable of recovery from damage. As the nodules take millions of years to grow the effects of deep-sea mining of this resource may go way beyond the length of time humans have been on the Earth. The region most likely to be exploited first for manganese nodules is the Clarion-Clipperton Fracture Zone between Hawaii and Mexico, covering about 5.2 million square kilometres of which 4.2 million are of commercial interest. This is a larger area than that of the 20 largest countries of the European Union combined.

To date, the development of deep-sea mining has been inhibited by economics and technological limitations. However, the development of technology for the offshore oil and gas industry and ocean cable laying has paved the way for deep-sea mining. As I sit writing these pages, giant machines belonging to a company called Nautilus, are being trialled in the ocean off Papua New Guinea with the intention of beginning the first commercial deep-sea mining operations in 2019. If successful, Nautilus will mine seabed massive sulphides from a deep-sea hydrothermal vent site 1,600 metres below the surface. Because the mining prospect is within the exclusive economic zone (EEZ) of Papua New Guinea, the company has received a licence to mine from the government. Exploratory licences have also been granted in the Solomon Islands, Tonga, Vanuatu and New Zealand. In the waters beyond national jurisdiction mining is regulated by a UN body, the International Seabed Authority. The Authority is based in Kingston, Jamaica, and I have visited it several times for discussions on environmental issues related to potential deep-sea mining in what is called the Area, the seabed that lies beyond national jurisdiction, below the high seas. Thus the water column of the high seas and the seabed (the Area) are, surprisingly, under different legal regimes. One hundred and sixty-eight states are officially members of the ISA, including China, India, Japan, the Republic of Korea and the United Kingdom. In the concrete buildings of the ISA these nations are getting together on an annual basis to decide upon the rules for deep-sea mining in the Area and carving up the seabed in the form of licences to explore for minerals. This might sound like a form of appropriation, but of course we are assured of the freedom to navigate and to undertake scientific research in the ocean which lies beyond state waters. However, this has not been my experience.

In late November 2011, I was over part of the Southwest Indian Ridge on the RRS *James Cook*. I had led the work on the seamounts of the ridge, but another scientist, John Copley, who had been a participant on the cruises in which we found and surveyed the Antarctic vents, took over as Chief Scientist to survey the Dragon hydrothermal vent field. Dragon, at the time, was the only area of deep-sea hydrothermal venting that had been located with certainty on the Southwest Indian Ridge, and it lay at a depth of between 2,700 and 2,800 metres in very broken terrain. *Kiel 6000*, the ROV, soon revealed the alien beauty of these unusual hydrothermal vents. While some of the vents were not much more than heaps of black sulphide draped in rust-coloured silt others resembled slender cones, in places a dirty dark turquoise or cobalt blue, with patches of deep scarlet and sulphurous yellow. Slender structures, looking like the antlers of a stag, grew from the summit of one chimney as well as sprouting from its sides. This received the name of 'Jabberwocky' from one of the scientists on board. Other vents were given names like Knucker's Gaff, Tiamat, Hydra and the less monster-like Jiaolong's Palace. In clusters around the bases of the chimneys were large black snails, very similar to the ones we had described from the Antarctic, clumps of giant, rust-covered mussels and the odd pale blue-white shrimp crawling here and there. John was in the ROV container directing operations while I sat with other scientists around a large screen in the plot room getting live footage from the cameras. Suddenly, among the large black snails, and being visibly barged about, was a small yeti crab, also peppered with rust but unmistakable. I almost fell off my chair in surprise and scrambled to grab the phone.

'John, John, get them to back the ROV up. I'm sure I've just seen a yeti crab in among the snails.'

There was some mumbling in the background as John gave instructions to the ROV pilot. As the camera tracked back, sure enough there were our furry friends from the East Scotia Ridge, scuffling for space with the black-shelled molluscs, which were threatening to overwhelm them. This was an animal I was not expecting to see in the Indian Ocean, and their presence indicated a biological link between the East Scotia Ridge in the Antarctic and the Southwest Indian Ridge.

'Yes I see them!' came John's happy voice from the ROV control van.

We set about collecting animals with the slurp gun on *Kiel 6000*, a device like a giant hoover, which we had frequently used on the vents in the Southern Ocean. Soon we were all in the cold-temperature laboratory, stooped over trays of the samples retrieved from the seabed deep below.

The yeti crabs were almost identical to the animals we had collected from the Southwest Indian Ridge, but were much smaller, probably less than half the size. The giant snails comprised two species. One had a blue-black shell with a pale pinkish foot. The other looked superficially similar, but the foot was covered in black scales. This was the famous scaly-foot snail, a local speciality only found on the deep-sea hydrothermal vents of the Central and Southwest Indian Ridges and the only animal to secrete iron armour as a form of defence. The scales covering the foot resembled small black tiles and were impregnated with iron sulphides, and the shell was also coated in this material. Scientists had known about this snail for 10 years and details of it had been published in a paper in 2003. However, for a reason which was not obvious to us, the authors of the paper had failed to give it a Latin name and also had not deposited type specimens in a museum. My PhD student Chong Chen stood leaning over the snails, which were about the size of a child's fist, completely mesmerized. Chong, who was from Hong Kong, had collected shells since childhood. Even as an undergraduate on the Biological Sciences degree in Oxford, his knowledge of molluscs was prodigious. He had a naturally happy-looking face and his look right then was just pure delight, a pleasure to behold.

The animals were truly remarkable and were, indeed, quite beautiful. It is thought that the iron is secreted for defensive purposes, probably against predators but also the hostile environment around vents, which includes hot, acidic vent fluid, hydrogen sulphide and heavy metals, all of which are poisonous. We set about preserving some of the precious samples to take back to Oxford. John completed his surveys of the Dragon, or Longqi, vent field as it is known in Mandarin. There was a serious purpose to this part of the expedition. Longqi had been licensed for exploration for deep-sea minerals by the Chinese. John was keen to establish a baseline of the site before any activities related to mineral prospecting began.

Five years later, in 2016, I was aboard the Japanese research vessel *Yokosuka* with the intention of returning to the Longqi hydrothermal vent field. This was with the intention of surveying the site again so that we could assess the impacts of exploratory mining activities undertaken by the Chinese. Chong, my student who had been so delighted in the finding of the scaly-foot snail was now a research fellow of the Japan Agency for Marine-Earth Science and Technology (JAMSTEC). He had described the snail and I had compiled his photographs into an anatomical drawing, which took several weeks of work to produce. A paper naming the species officially for the first time was published with the permission of the Swedish scientist, Anders Warén, who had undertaken the preliminary description of the species. It was extremely gratifying to see the work come together. The scaly-foot snail had turned out to be a remarkable beast. It had a gland in the foot packed with symbiotic bacteria that oxidized hydrogen sulphide from the vent fluid to provide the energy to grow and provide the snail with food. As a result, the scaly-foot had a reduced gut size compared to related snails because of the lack of a need to feed. The animal had also evolved to cope with the low-oxygen environment of the hydrothermal vents with a gill that extended down most of the body length and a sophisticated blood circulatory system. This included a massive heart, which made up about 4% of the body volume compared to a human's heart which makes up about 1.3% of the body volume. Chong had also undertaken genetic investigations of the scaly-foot snails from the Longqi hydrothermal vent field compared to populations living around vents on the Central Indian Ridge, lying to the north-east of the Southwest Indian Ridge. These studies showed that the Longqi vent populations were genetically different from those on the Central Indian Ridge, an indication that population connectivity was very poor between the areas. This was significant, as should a catastrophe strike the Longqi vent field, killing all the scaly-foot snails, recolonization from the Central Indian Ridge populations, via larval dispersal, was unlikely. Natural catastrophes, such as volcanic eruptions, did strike deep-sea hydrothermal vents, although these were probably less frequent on slow-spreading ridges such as the Southwest Indian Ridge than in settings where ridge spreading rates were much higher. However, our concern was the fact that the Longqi vent site had been licensed to the Chinese for mineral exploration.

While the Indian Ocean sounds idyllic, we spent the first couple of weeks sailing through intermittent poor weather as a result of the proximity of a cyclone. The organization of a Japanese ship is much more hierarchical than on a British vessel, with communication between the ship's officers and the scientists strictly through the Chief Scientist, in this case a lively character called Ken Takai. Chong in turn relayed news to the British scientists on board, Nick Roterman, a veteran of the Southern Ocean vents cruise, and me. The Japanese had informed the Chinese government of an intention to work at the Longqi vent site, but apparently there had been no response. There had been two Chinese research vessels in Port Louis in Mauritius and communication with the Chief Scientist on the Chinese expedition had met with a rather unreasonable demand for a third of all the specimens collected by us should we visit the Longqi vent field. It is important to remember here that scientific research is one of the fundamental freedoms of the high seas under the UN Convention on the Law of the Sea. Despite the existence of a licence for the Chinese to explore the area for minerals, there were no grounds to refuse us permission to do research in the area or to restrict it. The demand by the Chinese scientist was turned down, and we anxiously waited on the ship for the next move. The response came several days later. Allegedly it had been stated that we were free to undertake submersible dives on the Dragon vent site but that we would risk a collision with an autonomous underwater vehicle, a robot packed with scientific instruments that resembles a torpedo, which would be operating over the site. Ken had no choice but to turn away from the Longqi vent site and head for the hydrothermal vents on the Central Indian Ridge. Needless to say we were bitterly disappointed and angry. I was also suspicious that perhaps we were being obstructed because of damage that had been done to the pristine site we had visited in 2011, five years before.

At the time of writing this, the International Seabed Authority has granted 29 licences for exploration for mineral resources in areas beyond national jurisdiction. The companies that have received these licences, many of them state-owned, are under obligation to undertake environmental studies as part of the licence conditions. However, the science that has been carried out has been of variable quality and is insufficient to understand the likely impacts of mining on deep-sea eco-systems. It should be borne in mind here that the deep sea is the largest ecosystem on Earth and also the least explored and understood. Mining of seabed massive sulphides and cobalt crusts will involve placing huge excavation machines on the sea-floor that break up and crush the seabed rock and then pump it to a barge on the ocean surface. Any animal living on the seabed in the path of these mechanical behemoths will be obliterated. A plume of sediment will be suspended in the waters over the mining site and added to by waste material released from the surface barge back down to the seabed. These plumes are likely to include toxic materials, such as heavy metals and hydrogen sulphide, usually mostly confined to the vent site itself. The operations will be in an environment that is usually dark, and will therefore be intensely lit and generate a lot of noise, the latter potentially disturbing marine life such as whales, which rely on sound for navigation and communication. Manganese nodules, which lie in very fine sediments will effectively be ploughed up or hoovered up, again generating large clouds of suspended sediment and removing the nodules themselves, a habitat for animals which live attached to hard surfaces.

The question here is whether deep-sea mining will take the same path as other industrial activities in the ocean that have gone before. Will humankind make the same mistakes again? Whether it is sealing, whaling or deep-sea bottom fishing our record on sustainable exploitation of ocean resources is not good. The pattern is a familiar one: technology enables the exploitation of a new resource and before scientific investigation allows for understanding of the activity's impacts, exploitation is rapidly ramped up. Without the science, civil society has no opportunity to make an informed decision on whether such resources should be exploited, and if they are, how such activities should be regulated. Industry has shown time and time again that the pursuit of profit leads to destruction of species and the environment. Our relationship with the ocean has been an abusive one, and the signs for deep-sea mining are not good. The activity is very likely to be highly damaging. It is targeted at some of the ocean's most sensitive or rare ecosystems.

The recent granting of an exploration licence adjacent to the Lost City hydrothermal vent site on the northern Mid-Atlantic Ridge has caused particular alarm. As I described in Chapter 2, this site is an alkaline hydrothermal vent which is globally unique and is of major scientific interest as possibly having conditions that are close to those on Earth when life first emerged. Lost City may hold the secret to genesis. It also forms part of an Ecologically and Biologically Significant Area (EBSA), identified under a process through the Convention on Biological Diversity, which identifies ocean features that are of importance to marine biodiversity, ecosystem function and conservation.

The decision to allow exploration for minerals in close proximity to Lost City, among others, has given rise to concerns about transparency and methods of decision-making at the International Seabed Authority. It is also linked to one of the most worrying aspects of the potential 'gold rush' for ocean minerals, the lack of comprehensive strategic assessments of the effects of mining operations within any region. Such assessments are required not only to account for multiple mining operations, but also other human activities, such as fishing, shipping, etc., as well as evaluating areas of ecological and conservation importance. Also, as my experience in the Indian Ocean has demonstrated, there is little doubt that mining licences, whether for exploration or production, are viewed as a form of appropriation by at least some of the licensees. As you read these pages the common heritage of humankind is being carved up by states and industry greedy to seize the resources of the ocean.

The Kogi concept of the ocean as providing an endless bounty is true in many ways. The ocean can not only feed us, but it can also provide us with drugs with miraculous properties to fight everything from cancer to the common cold. It maintains the Earth in a way that allows humans and the rest of life to live comfortably. It may provide us with the minerals to drive a fourth industrial revolution and the means to move society away from dependence on energy provided by hydrocarbons. However, this will only remain the case if we carefully manage the ocean and limit our impacts on marine ecosystems. There is an old farming adage about sowing seeds which runs along the lines of: 'One for the mouse, one for the crow, one to rot and one to grow.'

For me, this saying speaks of leaving a part of the world for nature. As I hope I have shown in this book, setting aside portions of the ocean as marine resilience or recovery zones and carefully managing the rest is a way of ensuring that marine ecosystems continue to provide everything we, and the planet, require. One problem with this, though, is that there is currently no legal regime for the establishment of marine protected areas in waters beyond national jurisdiction. As I write the final words to this chapter, politicians, lawyers, scientists and members of civil society are debating a new implementing agreement for the UN Convention on the Law of the Sea in a series of intergovernmental conferences in New York. The focus of the new agreement is on the protection of biodiversity beyond national jurisdiction. This will include a new legal framework for the establishment of spatial conservation measures in waters beyond national jurisdiction, the setting of standards for assessment of environmental impacts of human activities in the ocean, and the fair and equitable sharing of the benefits from the genetic wealth of the ocean. It took many years to get states to agree to these negotiations, but the ocean's moment has arrived and they give me great hope that there will be a major step forward in our ability to sustainably manage human activities in *all* of the ocean.

# 10

# Conclusion:

## *The Choice Between Two Oceans*

I believe we stand at a critical moment in history. Scientists have observed, measured, surveyed and modelled to the point where we have a good understanding of how the ocean is dying. And it is dying. Unlike our ancestors, who had the excuse of ignorance, we know that we are changing the ocean dramatically, at an unprecedented rate and for the worse. From rising temperatures, acidification and deoxygenation, to plummeting populations of iconic marine species, to trash washed up on beaches, the warnings are loud and clear. Despite the alarming symptoms, however, as far as we are aware, extinctions of marine species have been relatively few compared to land. We still have everything to play for.

We therefore face a stark choice between two very different oceans. One is healthy and productive, where fishing, aquaculture, tourism and other industries are sustainably managed, and where we have stopped the activities that damage the ocean needlessly, such as dumping our waste in it. Such an ocean would demand that policymakers and society across the globe have the information they need at their fingertips to make the right decisions on where and how to protect the ocean and then to monitor how the ocean responds to our actions. It would not necessarily look like the ocean of the past, before industrial fishing and the ravages of climate change, but it would retain a high diversity of species and be healthy and productive. Such an ocean would support all the ecosystem services we rely on, including the obvious ones such as food, but also the ones we do not necessarily recognize such as atmospheric regulation, nutrient cycling and coastal protection. The alternative is to continue on the current trajectory of decline. The thousand cuts delivered through overfishing, destructive activities, marine litter, pollution and climate change will drive entire ecosystems, such as coral reefs, into collapse and species will disappear at a rate comparable to a historical mass extinction. Just at the point where we need the services the ocean provides to support a world of 10 billion people we will find that we have exhausted it. This will not only be a case of the slow degradation of critical ecosystem functions, but also of an increasing frequency of unpredictable environmental catastrophes as tipping points in the Earth system are crossed. We are already seeing such manifestations in the shape of more extreme weather, the outbreak of uncontrollable forest fires in places where they have never been seen, and the sudden destabilization of ecosystems like coral reefs. The parallel

# Conclusion

with Rachel Carson's two roads of the early 1960s, outlined in Chapter 6, is unavoidable.

Now, as then, we can simply bury our heads in the sand and continue with business as usual. Or, alternatively, we can stand up and demand that our governments do something about the situation and do something fast. As this book has moved from conception to words on a page, two thirds of the northern part of the Great Barrier Reef has died in two successive years of mass coral bleaching driven by the hottest years on record in that region. Scientific evidence of the impacts of climate change, overfishing, habitat destruction, pollution, decline of our iconic ocean predators and the effects of invasive species, arrives daily. Among these the prognosis regarding climate change is particularly alarming. That time is running out rapidly is very clear to the scientific community and more enlightened governments and members of civil society. So what is holding us back from reaching for solutions? I could write another book on the reasons for lack of action, but in my opinion, several factors have been very important over recent years.

The rise of nationalism and the spurning of international law and cooperation has to be a strong contender for why humanity is currently in paralysis in the face of this shared battle. The priority of self over the greater good underlies many of the problems we see in ocean governance and management today. It is driving appropriation of ocean resources that belong to all of us. This has been seen in arguments over extending the jurisdiction of states over more continental shelf and adjacent seabed so that they gain the rights to the resources of the seabed. The annexation of reefs in the Spratly Islands by China, and to a lesser extent, by the Philippines, Vietnam, Malaysia and Taiwan, followed by land reclamation and building of military bases, is a good example. Despite such action being against international law and also illegally destroying large areas of healthy coral reef, states eager for fish, oil and gas resources in the region have gone ahead with claiming the islands despite international protest. The alternative, whereby nations in the region could have come together, suspended their territorial claims and agreed to jointly exploit resources in the region, seems to have been barely attempted.

Elsewhere, including in the Arctic, and the eastern Mediterranean and Aegean Seas, disputes are also active, although not as tense as the current situation in the South China Sea. In almost all cases these disagreements relate to discoveries of oil and gas, which, given the situation with global climate change, should be left under the seabed in any case. In countries with weak governance, appropriation by industry, whether it be mining, agriculture or oil and gas, has taken a very sinister turn. According to the *Guardian*, in 2017 over 200 environmental activists were murdered in the space of a year for trying to defend ecosystems or species from destruction. Many of these people were from indigenous communities. As the struggle for ever more scarce resources has intensified, individual citizens have found themselves in the frontline, often criminalized or killed with impunity. This is another symptom of the polarization between those who would extract the Earth's resources regardless of the cost to the majority over those who believe in sustaining the rapidly dwindling natural environment. It also manifests when wealthy states are the only ones able to take advantage of resources in the ocean. This is certainly the case for some high seas fisheries that demand expensive and highly sophisticated fishing vessels to participate. Another example that is emerging is the ability by wealthy countries or companies to investigate marine life for natural products that may lead to the discovery of new drugs or other chemicals of commercial value. As demonstrated by BASF, they do not even have to sample the ocean, just screen DNA databases that are publicly available. Marine mining in areas beyond national jurisdiction is also likely to go the same way although there are requirements for benefit sharing among states for this activity. How that will work in practical terms is yet to be seen. Appropriation does not need to be the drawing of lines around maps to lay claim to parts of the ocean.

Nationalism has manifested in other ways in the international system of ocean governance. Usually in the name of protecting their (self) interests, states have consistently worked to weaken and undermine international regulations relating to the ocean. Despite the UN Convention on the Law of the Sea being agreed in 1982 and coming into force in 1994, its implementation remains inadequate. Various UN bodies are charged with implementing the Law of the Sea Convention. They are chronically underfunded, but also largely deal with single sectors of activity. Examples include the International Maritime Organization which deals with shipping and safety at sea, the UN Food and Agricultural Organization dealing with fishing and aquaculture, and the International Seabed Authority dealing with mining. These organizations are very protective of their respective charges, meaning that looking at management of human impacts on the ocean as a whole becomes extremely difficult. Decision-making within these organizations has often lacked transparency and therefore accountability. The decision-making structures and processes within these institutions have also encouraged the lowest common denominator in terms of what states collectively are prepared to agree to. An example is provided by many of the regional fisheries management organizations (RFMOs) that are charged with the management of some of the major global fisheries, such as for tuna, and also fisheries beyond national jurisdiction. Furthermore, especially with respect to fishing, exceptions have often been made to protect the interests of states and industries rather than the ocean. The historical exemption of fishing vessels from requirements to display an International Maritime Organization registration number is one example where the only beneficiaries can be those who are out to break the rules. The latest development with the International Whaling Commission whereby Japan has simply decided to pull out because it objects to the international moratium on hunting whales is extremely worrying in this context. I also find it remarkable that the United States of America has never ratified the Law of the Sea Convention when clearly it benefits everyone, as imperfect as it is.

Climate change negotiations are perhaps where issues of nationalism have been most obviously destructive in recent years. Again and again, these have unravelled as states have fought over who should cut emissions and who is responsible for the damage to date. Rich developed nations refuse to take responsibility for what they have already dumped into the atmosphere and less developed nations will not accept emission reductions that threaten their economic growth. The recent lurch to right-wing nationalist governments in some of the major states of the world, such as the US and Brazil, political chaos arising from the economic crash of 2008, and political upheavals such as Brexit only distract from what is the most pressing issue of our time. I remember sitting in the UK Parliament at an all-party committee hearing evidence about the state of the ocean and being asked specifically about what the effects of climate change are on marine ecosystems. In response I discussed mass coral bleaching, ocean deoxygenation and acidification.

'Surely you are exaggerating?' claimed one Member of Parliament. 'These things are only predicted for the future.'

I had to explain to him that these were actually the things that have happened already, that the future predictions are much worse.

There has been a wilful dismissal of scientific observation and evidence related to climate change by politicians and industry with deeply held political convictions or vested interests. While the lobbying by the oil and gas industry against climate change action in the early days of the Intergovernmental Panel on Climate Change (IPCC) has been clearly identified, it is important to realize that it is still an active force today. In almost all such cases I point to the ocean and ask people to simply look and see what is already in progress now. As the preceding chapters of this book have identified, ocean warming, acidification and deoxygenation are real. They are already having catastrophic impacts on marine ecosystems and the predictions are that, at the current rate of $CO_2$ emissions into the atmosphere, things are going to get much worse, very quickly. This is nothing less than a global emergency.

The other major issue in tackling problems in the ocean is simply a lack of awareness. As a child, the most important part of my holidays was the time I spent by myself on the beach, clambering around the rocks of the bay in County Sligo, Ireland. I spent hour after hour and day after day on these rocks, searching in small rock pools or lifting up the seaweed or boulders to see what was hiding underneath. It was here that I began to meet the incredible variety of marine life, and my obsession with the ocean was sparked. As time went on I collected guides to the seashore. *The Hamlyn Guide* by Andrew Campbell, with whom I would work on a deep-sea expedition off Oman many years later, was my favourite. This chunky book was illustrated with paintings, several to each page, of everything from seaweeds to snails and fish. I used to skim these books all the time, sponging up knowledge as children tend to do. Thinking back, though, one of the fascinating things about my weeks in Ireland was the response of my grandfather to these books. He was as tough a fisherman as they come, but he would sit by the fire in the front room of the cottage with these books, thick-rimmed glasses perched on his nose. He would point out things he had seen in the lobster pots or raise his eyebrow at a fact about some 'queer buck' or other, referring to a creature. This fascination and wonder I saw in him was in sharp contrast to the brutality that I would sometimes see him visit upon the marine life on his boat. I often wondered, looking at him, if he had known more about the incredible world he relied on for his livelihood, whether he would have more valued the animals he encountered on a daily basis as part of the same ecosystem that provided him with the lobster and crab.

Considering my own grandfather's lack of knowledge about the marine life upon which he relied for a living, I don't think it is at all surprising that as adults, we simply do not register the ocean's importance. The ocean is distinctly absent from the school curriculum, from nursery to senior school. Many people only come into contact with it when eating the odd fish caught from the sea – fish and chips if you are in the UK – or from lying on a beach or snorkelling while on holiday. Strangely, even a highly damaged ocean can appear healthy to someone who doesn't know what it looked like in the past. Why should they know the fish or the corals are missing if they didn't know they were there in the first place? This is well known in the scientific literature and is called the 'shifting baselines syndrome'.

Alarming stories appear in the press with regularity, whether it is about ocean plastic, dying coral reefs or stranded whales. These either appear so briefly that while they are sad stories, they hardly seem to matter, or the public become habituated to hearing the bad news and switch off as they feel powerless to help the situation. The BBC's *Blue Planet II* has been an exception, through awakening the public and politicians to the issue of marine plastics. The beauty of the series' first episodes was matched by some of the horrors of its final episode. In the UK, the series has galvanized government and the public to try and stop the use of plastic bags, bottles and drinking straws everywhere. Why was this message successful when so many others have not been heard? I think it is because the use of disposable plastics is something we all contribute to. I can choose to buy a reusable steel bottle for my water or I can choose not to give my children plastic straws for their drinks.

Ignorance is inexcusable. Politicians have a responsibility to safeguard the environment for their voters and for future generations. It is their duty to understand what the problems are and to try and solve them. Corporations and financial institutions have grown larger and more powerful, and in some cases are effectively wealthier than individual countries. With this wealth comes responsibility. It is no longer acceptable that industry simply makes profit at cost to the environment, and by extension to the rest of humanity. Industry must take some of the responsibility, especially when government budgets may be very stretched, to look after the environment and should certainly not deliberately damage it.

Our lack of knowledge of the ocean's ecosystems and behaviour contributes to our ignorance of its role in our world. As I mentioned earlier in the book, though it may seem a remarkable figure, we have probably only investigated 0.0001% of the deep-sea floor and even less of the deep-water column lying above it. For seamounts we have explored only a few tens out of more than 170,000, so around 0.02%. For ocean trenches and canyons our knowledge is even thinner. I have seen various estimates for the number of species that live in our ocean from 300,000 to two million. Given the rate of discovery of new species of larger organisms I have witnessed on my own expeditions, I feel that even the higher figure is probably an underestimate.

# Conclusion

On the recent Nekton Foundation expedition to Bermuda I undertook as Chief Scientist, we uncovered more than 40 new species of seaweeds and numerous new animals, including corals and various crustaceans. Every species of large animal we found on the deep-sea hydrothermal vents of the East Scotia Ridge in the Southern Ocean was new to science. This was not just at the level of species, there were new genera and even new families of animals as well. Suffice it to say we do not fully understand how much life is in the ocean, how many species there are and what their geographic range might be. Currently, scientists cannot, with any reasonable accuracy, estimate which species can be found where. If we do not fully understand how life is distributed in the ocean, we certainly do not have a firm grip on how species function within ecosystems and, in turn, how they might benefit the planet, and by extension, us. Without such knowledge, making evidence- or science-based decisions on how to sustainably manage the ocean is incredibly difficult. But by the same token, a lack of knowledge should not prevent us from making some important decisions. It simply means that in many cases these decisions have to be highly precautionary.

So, what needs to happen for us to choose the right road and rejuvenate the ocean?

**1. Nations must reach and implement an effective agreement to immediately slash $CO_2$ emissions as quickly as possible.** This may sound boring but if we do not decarbonize the global economy, ensuing climate change will continue to destroy marine ecosystems, impacting biodiversity and significantly reducing the services provided to humanity by the natural environment. Even to limit global atmospheric warming to below 2°C, the currently agreed target of the Paris (climate) Agreement, will still be highly damaging to the ocean. The Independent Panel on Climate Change is now recommending a more aggressive programme of cuts to $CO_2$ emissions to prevent atmospheric warming from exceeding 1.5°C to prevent catastrophic damage to the environment. Ultimately $CO_2$ emissions are a threat to human civilization, as we are dependent on this very natural environment to underpin our survival on Earth for everything from food and water, to a stable climate. Action is required in all human activities that generate greenhouse gases, including in transport, agriculture, construction, industry, domestic energy usage and even our diets. Humanity has to wean itself off fossil fuels.

**2. Eliminate overfishing and, where technically possible, destructive fishing practices.** Although many fisheries in regions controlled by highly developed countries have stabilized, overfishing is still rampant over much of the ocean. Yet with our current understanding of how to sustainably manage fish stocks, and sophisticated technologies enabling the tracking of fishing vessels and tracing the catch from the sea to supermarket shelves, this is a problem that we can, with effort, solve relatively easily. I once heard a famous fisheries scientist, Daniel Pauly from the University of British Columbia, say of the current state of fisheries management: 'It is like having a patient in the most modern of hospitals surrounded by state-of-the-art scanners, technical equipment and well-trained doctors. But

Conclusion

nobody lifts a finger to treat the patient. They are just left there to die on the trolley.'

The whole fishing system – from extracting fish from the ocean through to the supply of fish to markets – must be entirely traceable, with no room for illegally caught fish. The technology is there to track vessels: satellite systems for the larger trawlers, purse seiners and long liners; mobile phone technologies for inshore fishing fleets. The new Port State Measures Agreement aimed at preventing landing of illegally caught fish at ports needs to be widely adopted internationally. All fisheries, from industrial fishing fleets to smaller artisanal fisheries, must be regulated and controlled. This means that every part of the ocean should be covered by appropriate fisheries management agreements, and measures and catches should be commensurate with sustaining stocks at levels where they can safely replace themselves into the future. This will mean painful action to reduce fleets where there is overcapacity and to remove harmful fisheries subsidies that artificially support oversized fishing fleets or unprofitable fishing operations. It will also mean cooperation between countries on a regional basis, to harmonize fisheries management in coastal waters and offshore in the high seas. The global community will also have to solve current issues around jurisdiction, flags of convenience and tracing of vessel ownership. Under current international law the best way for this to be dealt with is by cooperation between countries.

**3. Establish a global network of effectively enforced marine recovery and resilience zones.** Studies have clearly demonstrated the benefits of marine protected areas or marine conservation zones in leading to the recovery of the abundance and diversity of marine life and to improve the resilience of ecosystems against shocks. These benefits extend to commercial species, as discussed in Chapter 8. Currently, just over 7% of the ocean has been protected, and the current international target is 10% of ocean area. However, studies of the benefits of marine protected areas suggest this target should be 30% and there are already calls from scientists, non-governmental organizations and intergovernmental organizations for 30% by 2030. Science tells us that the most effective marine protected areas are old, large, have no fishing in them, are well-enforced and are in areas that are relatively isolated from human influence. Not all conservation zones have to be large, though, as smaller areas can be effective, especially if they are protecting specific features or habitat. Presently, many protected areas allow fishing and many are poorly managed, rendering them less effective, or at worst, ineffective.

**4. Effectively reduce the problem of marine pollution.** This requires a range of measures, many of them on land, to prevent harmful chemicals and debris from reaching the ocean. We are yet to fully understand the implications for marine life of the contamination of the entire ocean with plastic. Science has to play catch-up to identify just how hazardous plastic waste is to the smallest forms of life right through to the largest animals and entire ecosystems. Single-use plastics must be eliminated from society and replaced with materials that are easy to recycle and which degrade completely in the environment. For many countries, investment in solid waste management is essential. The issues around persistent and pseudo-persistent chemicals are even more insidious as they cannot be seen in the environment and their biological effects can be difficult to resolve. Here, the only solution is reversal of the burden of proof on industry. Manufacturers of chemicals that are finding their way into everyday household products must demonstrate that the materials being sold do minimal harm to marine ecosystems. The responsibility does not remain with the producers alone, though. Companies adding these chemicals to their products must also show a duty of care to the environment. Some chemicals, such as many medicines, cannot be banned and so better water-treatment technologies are needed to try to remove harmful substances from waste water before it flows into rivers, and from there to the ocean. Technical innovations that remove or prevent waste from entering the oceans are needed. These represent significant business opportunities for companies and countries willing to invest in them.

**5. Improve management of the ocean.** Individually and collectively, the national and global institutions currently charged with managing various aspects of human activities that impact the ocean are weak. Cooperation across sectoral and institutional boundaries must dramatically improve if we are to meet the current challenges leading to ocean degradation. This is probably best achieved on a regional basis where countries and respective institutions can get together and resolve issues in ocean management in a more holistic fashion. Only through such cooperation can the strategic assessments be made to identify current impacts on the ocean and to make informed decisions to ensure that exploitation and development is sustainable.

Along with improved cooperation in governance must be a new era in terms of transparency. Civil society must have oversight of how and why decisions are made in such national and international governing bodies. Without absolute transparency such oversight is not possible. Experience has demonstrated time and time again that when decisions by implementing agencies take place behind closed doors, weak decisions and influence from external lobbying – usually by industry with vested interests – abound. It should not be acceptable, in a world where fishing is sustainably managed, for data on where fish are caught and in what quantities, to be confidential. It should also not be acceptable for fisheries management organizations to ignore scientific advice on sustainable catch levels by setting quotas too high. They must be subject to scrutiny and questioning by civil society. I pick on fishing here as an example, but such transparency should apply to all industries that use or rely on the ocean.

**6. Take greater steps to map the ocean and the distribution of life within it, and to understand how it works.** Levels of funding for marine science and the mechanisms by which it is undertaken are inadequate to gather the information we urgently need to understand the ocean and improve its management. Traditionally, scientific projects are funded in three to five year chunks and run by individual principal scientists or, at best, small teams of collaborators. Much more ambitious missions are required, covering large geographic areas and lasting decades rather than a few years. It always amazes me how much money is regularly spent on space projects, aimed at exploring planets within the solar system or even distant stars and galaxies. Such projects have cost billions of dollars and can last for 30 years or more. Yet only tiny fractions of our own inner space, the ocean, have been explored. Given the importance of the ocean's resources to humankind and its critical role in how the Earth is responding to human $CO_2$ emissions, it is time we took steps to change our understanding of the ocean. Such a new wave of ocean science and exploration will also require new ways to store, access and use ocean data so that all parts of human society can benefit from it, particularly through its use in decision-making.

**7. Increase efforts to educate all sectors of society on the ocean and its importance.** This is especially true for schoolchildren, the next generation, for whom the ocean can be key to unlocking interest, whether it be in the natural world or in technology and engineering. Without an appreciation of the ocean at an early age, how can we expect the politicians, business leaders and scientists of the future to help conserve the ocean and use it in a sustainable manner?

So these are all the big ideas that need international cooperation and vision. You might be forgiven as an isolated reader for thinking there is little you can do to help the current plight of the ocean. Many times in the past when I have spoken publicly about ocean degradation, people have asked what they can possibly do. For me the answer is a simple one: every single small thing you can do helps the situation by reducing some small fraction of the current impacts on the ocean or by helping it to recover. So here they are, my tips for a personal mission to shift the oceans from decline to recovery:

**1. Reduce your carbon footprint.** Economize where you can on energy consumption at home. Avoid using the car where possible – cycle, walk to work or use public transport – it can even improve your health! Think about the need for flying, and if you must fly, can you offset your carbon footprint as an individual or as a company? Reduce the food miles of the items you buy in the supermarket and consider how carbon friendly your diet is.

**2. Make yourself aware of what fish are considered sustainable.** Various organizations produce guides to the sustainability of fish and shellfish. An example in the UK is the Marine Conservation Society's Good Fish Guide (mcsuk.org/goodfishguide/search). In the USA there is the Monterey Bay Aquarium's Seafood Watch which is available as a mobile phone app. The World Wildlife Fund produce a list of different sustainable fish guides for different countries (wwf.panda.org/get_involved/live_green/ out_shopping/seafood_guides). Many of the products for sale in supermarkets now, at least in the United Kingdom, have various forms of certification. It is important to check on what these actually mean, however. Considerations of sustainability should also extend to aquaculture products.

340

# Conclusion

**3. Avoid single-use plastics.** All those disposable plastic-bottled drinks, plastic bags, plastic straws, plastic plates, plastic cups and no doubt many other disposable items. Choose items that are reusable (e.g. steel water bottles, reusable cloth bags), or if you must use disposable items, make sure they are recyclable or compostable. Believe it or not, even the clothes you wear make a difference to the pollution of the ocean. Artificial materials, such as polyester, release tiny microfibres when they are washed which flow into our water treatment systems and find their way to the sea where they accumulate. Cotton and wool are natural fibres that are broken down in the environment and are much more ocean friendly.

You might be astounded at how much plastic waste you can avoid producing as an individual and especially as a business.

**4. Be aware of the chemical content of everyday household items you use.** No doubt when you look at the list of ingredients on a bottle of shampoo or shower gel you are faced with a large list of chemicals you might be unfamiliar with and therefore will probably think this is useless advice. Do not despair, however. Help is at hand. Consumer information is available from various sources, and one I have found useful is the website of the Environmental Working Group (ewg.org). This can certainly be informative in terms of identifying the types of products and chemicals you need to consider when buying household and personal care products. Obviously, I do not want to frighten everyone to death and stop people using products that may offer protection or benefits to their well-being or health. Judgement is

341

required in deciding whether or not you should use a product or find an alternative that is more healthy for the ocean.

**5. Educate yourself about the ocean, its biodiversity, its importance to us and our impact as humans on it.** Reading relevant books (you've already taken this step if you are reading this), watching documentaries or researching on the internet are all fun ways to do this. Why not actually go to the ocean as an individual or with friends and family and spend a day with a guidebook trying to learn more about the marine life of your own coast or where you are spending time on holiday? If you are a parent, teach your child about the ocean, and if you are a teacher why not develop lessons that use the ocean to set relevant examples or create scenarios for a particular topic? Join an environmental organization and receive news and updates about ocean issues. You may be inspired to become a marine conservation volunteer or you might even decide to write to your Member of Parliament or other governmental representative.

**6. Consider the impact of leisure activities on the ocean.** Many of you will spend time around the beach or in the sea sailing, kayaking, scuba diving or fishing. Can you pursue your activity in a way that is less harmful to your surroundings? You might do so through using fixed anchor points, not touching or sitting on marine life when diving, not leaving your trash behind on the beach. Is the leisure company you are spending your money with behaving in an environmentally responsible manner? Even better, can you collect ocean data as a citizen scientist? There are

# Conclusion

more and more opportunities to do just that and become your own marine scientist.

I leave you here with a final thought. It was 1993 and I was aboard the RRS *Discovery* off the coast of Mauritania on my first deep-sea expedition. For me, the cruise was a revelation, uncovering for the first time the marvellous animals of the twilight zone: devil anglerfish, viperfish, fantastic red shrimps and bright purple jellyfish. One evening I walked up to the bows of the ship at midnight trying to escape the claustrophobic conditions and the noise of engines within the ship. It was a perfect starlit evening. As the ship gently rose and fell, cutting through the glassy waters of the Atlantic, the air blew past me in time like a cool breath. I stood taking in a star-filled sky like none I'd ever seen on land. Then from either side of the ship, out of the darkness, bright-green bioluminescent trails cut through the inky black waters and converged on the bow of the ship. I leaned over to see two dolphin shapes perfectly outlined in blue-green light just in front of the bows and underneath the surface. Every few seconds the dolphins would leap out of the water and blow as they surfed the pressure wave in front of the ship. They dodged across and back, changing places or zipping almost underneath the bows. I was utterly mesmerized, transfixed by a sight of ethereal beauty so unworldly that it simply took my breath away. Finally, together they peeled off and shot away to starboard like a pair of submarine shooting stars. They left me standing alone on that starry night with tears in my eyes and a memory burnt in my mind forever for which I will always be grateful.

# The Deep

There is nothing else like the ocean. It is capable of giving moments that in a second will sweep you off your feet and change your life forever. A majesty and a spirit that captures the essence of the wild, of life itself. We cannot let it die, for without it the human experience will never be the same again and we will almost certainly perish alongside the coral reefs, sharks, tuna, whales and even the life that dwells below in the darkness of the deep ocean. We have the opportunity right now to head off an ocean of shame and give to future generations an ocean of hope. Much of the ocean's biodiversity is still with us, we know it has the capacity for miraculous recovery, and it can continue to provide for us and for all life on Earth. By taking the actions I have outlined with extreme urgency, all of us, from you and me as individuals to politicians and governments, can save the ocean and the life it cradles, and at the same time guarantee a safer future for humankind.

# Postscript

*The dark oceans were the womb of life: from the protecting oceans life emerged. We still bear in our bodies – in our blood, in the salty bitterness of our tears – the marks of this remote past. Retracing the past, man, the present dominator of the emerged earth, is now returning to the ocean depths. His penetration of the deep could mark the beginning of the end for man, and indeed for life as we know it on this earth: it could also be a unique opportunity to lay solid foundations for a peaceful and increasingly prosperous future for all peoples.*

Arvid Pardo, Maltese Representative
to the UN General Assembly, 1967

# Glossary

Abyssal plain    The vast areas of the deep seabed that generally show low relief and are largely composed of fine muds and oozes. Generally defined as lying between 3,000 – 6,000m depth. In areas host manganese nodules.

Acidification    The lowering of the pH of seawater as a result of absorbing $CO_2$ which is then converted to carbonic acid. Lowers the concentration of calcium carbonate in seawater. It is an effect of climate change.

Algal bloom    A very rapid increase in the growth of microscopic algae in the surface layers of the ocean. Can also refer to cyanobacteria. Often identified by discolouration of the water (red or green).

Anoxia    Absence of oxygen.

Anthropocene    The age when human activities have become

a significant influence on the Earth's climate and environment.

**Archaea**
A domain of single-celled microorganisms of a similar size to bacteria. Like the latter, Archaea lack a nucleus and are therefore classified as prokaryotes. Eukaryotes, organisms like us, have cells with a nucleus, a distinct structure housing DNA and form a third domain of life (bacteria are a second domain). Archaea are distinct from bacteria in the structure of their cell membranes, cell walls and DNA processing. They were first found in extreme environments (e.g. hydrothermal vents) but are now known to occur more widely.

**Basalt**
A dark volcanic rock formed by the rapid cooling of magnesium and iron-rich lava.

**Biodiversity**
The variety of life on Earth encompassing variation at the genetic level of individuals and populations, through species and up to habitats and ecosystems.

**Bioluminescence**
The biological production of light. This usually involves a light emitting molecule (luciferin) and an enzyme (luciferase). Bioluminescence is very common in the twilight or mesopelagic zone where it can be used in signalling, to illuminate prey or predators, to lure prey, to dazzle predators or prey and as a form of camouflage.

**Biosphere**
The parts of the Earth where life exists.

**By-catch**
Species which are accidentally caught (and

often killed) when fishing for a specific species or range of species.

| | |
|---|---|
| Cell membrane | An envelope or coating of a cell formed by lipids (fats) and proteins. |
| Chemosynthesis | A process whereby chemicals are oxidised to release energy which is then used to transform carbon dioxide or other carbon sources to the organic molecules which make up life. This differs from photosynthesis which uses light as an energy source. |
| Clathrate | A gas, typically methane frozen in a cage of water molecules under low temperatures and high pressures often under the seafloor on the edges of continents. The clathrate or methane hydrate appears as a white ice-like substance that when exposed to air can be lit and will burn. |
| Clay minerals | Hydrated aluminium silicates sometimes with other metals associated with them. Can be fine-grained and include minerals such as kaolinite and are often found in soils. Associated with some theories related to the genesis of life on Earth. |
| Continental shelf | The area around continental landmasses submerged beneath shallow seas down to around 200m depth (deeper around Antarctica). |
| Cosmeceuticals | Cosmetics that are reported to have benefits to health. |
| Ecosystem | A community of living organisms interacting with their environment including the non-living components forming a system. |

| | |
|---|---|
| El Niño | A large-scale climatic phenomenon caused by the build up of warm waters in the western Pacific which then spill across the ocean to the eastern Pacific. Causes the usually cold waters associated with the Humboldt current to warm, killing marine life and also associated with torrential rains in South America. Influences global climatic variation. |
| Epifauna | Animals that live on the surface of the seabed. |
| Eutrophication | Enrichment of seawater with high levels of nutrients, usually nitrates and phosphates which then cause an algal bloom. |
| Euxinia | A condition where the ocean is lacking oxygen causing a build-up of free hydrogen sulphide poisonous to most marine life. |
| Foraminiferan | A type of amoeba that usually grows a shell made of calcium carbonate, silica or grains of sand or other material. Naked forms lack the shell. |
| Gaia | A theory that living organisms interact with the non-living parts of the Earth to form a self-regulating and complex system maintaining conditions for life to continue existing. First put forward by the scientist James Lovelock. |
| Goldilocks zone | The zone around a star in which it is possible for life to exist on a planet. |
| Greenhouse effect | The trapping of the heat of the sun in the lower atmosphere by certain gases such as carbon dioxide. This is a natural process which keeps Earth warm enough for life. The |

anthropogenic or human greenhouse effect has raised the concentration of greenhouse gases in the atmosphere as a result of human activities causing global temperature rise.

Hadean eon — Period between 4.6 to 4 billion years ago when the solar system was forming.

Hydrothermal vent — Hot spring on the deep seabed caused when seawater penetrates the (deep-sea) crust and comes into contact with hot rock, often associated with a magma chamber. The seawater is heated, becomes buoyant and rushes towards the seabed. Chemical reactions with hot rocks mean the water is enriched in chemicals such as hydrogen sulphide and metals such as copper but loses its oxygen.

Hypoxia — Conditions of reduced oxygen levels.

Hydrogen sulphide — A reduced chemical that is an important source of energy in microbial chemosynthesis. Highly toxic to most animals, including humans!

Krill — A small shrimp that can occur in swarms of millions individuals. Very important in some ecosystems such as the Southern Ocean where it is an important link in the food chain from phytoplankton to ocean predators such as whales.

Lava — Melted rock that is extruded onto the Earth's surface or onto the seabed.

Little Ice Age — A period of cool climate that occurred from the medieval times (after the medieval warm period) to around 1850.

| | |
|---|---|
| Lycopod | An ancient group of plants represented today by club mosses. They typically spread by producing spores, tiny structures that grow into new plants which can be resistant to adverse conditions. |
| Magma | Melted or semi-melted rock that occurs beneath the Earth's surface. |
| Medieval Warm Period | A period of warm climate in medieval times from about 950 to 1250. |
| Megafauna | Large marine animals visible in photographs or on video film taken from landers, towed cameras, ROVs or submersibles. |
| Mesopelagic | The zone between 200m and 1,000m depth where some sunlight is detectable but is insufficient for photosynthesis. Home to many animals adapted to life in dim light. |
| Mesophotic | Species or communities of animals that live in the low light zone below the depth of scuba diving. Mesophotic coral communities occur from ~30m to 165m depth and in the tropics comprise zooxanthellate or light-harvesting corals and other seabed life. |
| Microbiome | The community of microorganisms that live in or on a multicellular organism. Important to the health of sponges, corals and humans. |
| Micronekton | Small animals that have power enough to swim against ocean currents. Includes small fish, shrimps, oceanic squid, some gelatinous animals and other groups. |
| Microorganisms | Tiny organisms that can only be seen in a microscope. |

# Glossary

Mid-ocean ridge | Linear ridges on the ocean floor where magma wells up from deep below the Earth's crust forming new seabed. The ridges can be quite smooth or can be very mountainous with a central rift valley.

Neanderthals | Extinct species or sub-species of archaic humans that lived between 450,000 and 40,000 years ago.

Nekton | Large animals that are sufficiently strong swimmers to move against the direction of ocean currents. Include larger fish such as tuna and sharks.

Nutraceuticals | Product derived from food which is supposed to have additional health benefits, for example dietary supplements such as cod liver oil.

Omega-3 fatty acids | A group of fats which are essential to human health but which the human body cannot synthesise. We must therefore get these essential fatty acids from our diet.

Organophosphorus | Organic compounds containing phosphorus that have been widely used as pesticides. They are extremely toxic to humans and the group includes some nerve poisons used in chemical warfare.

Oxygen minimum zone | A zone of low oxygen below the surface layers of the ocean caused by the natural respiration of bacteria using oxygen to break down organic material sinking from the ocean surface.

Palaeocene-Eocene Thermal Maximum | An event 55 million years ago when global temperatures climbed 5-8°C over a period of less than 20,000 years. Associated with extinctions in the deep sea.

pH — A unit of measurement related to how acid or alkaline a liquid is. The scale is reciprocal and logarithmic so an acid has a low value and an alkali a high value. Hydrochloric acid has a value of 0.1, vinegar about 3, neutral tap water about 7, pre-industrial seawater about 8.2 and household bleach 12.

Photosynthesis — A process where light is used to provide energy to transform carbon dioxide into the organic molecules which make up life. Undertaken by plants, algae and cyanobacteria.

Phytoplankton — Algal cells that live in the upper sunlit waters of the ocean and which provide the base of the food chain for most marine ecosystems. They have no power of movement against ocean currents.

Plankton — Tiny organisms that live in the ocean but which have no power of movement against ocean currents.

Polymerization — A chemical process where small molecules (monomers) join together to form much larger molecules (polymers).

Protozoa — Single-celled eukaryotes (organisms with a nucleus where the DNA is stored).

Rare earth element — A group of elements that are dispersed in the Earth's crust and often occur with other minerals with similar properties making them difficult to extract. They are important in a number of important technologies such as

354

mobile phones, lasers, superconductors and magnets.

Rhodolith | Spherical limestone concretion made by marine algae.

ROV | Remotely operated vehicle. Robot tethered to a ship used to undertake survey and sampling in the ocean.

Sea-floor spreading | The process of formation of new seabed or oceanic crust at mid-ocean ridges.

Stratification | The formation of layers of water in the ocean usually driven by differences in temperature and salinity. It reduces mixing of water from different depths.

Symbiosis | A relationship between two species where both benefit and on which both are highly dependent.

Thermocline | A zone moving down from the surface of the ocean where temperature rapidly changes, usually cooling.

Thermohaline circulation | The large-scale flow of water through the oceans which is largely driven by differences in temperature and salinity.

Zooplankton | Tiny animals that live in the ocean and have no power of movement against ocean currents.

Zooxanthellae | The microscopic symbiotic algae that live in the tissues of shallow-water reef forming corals and some other animals. Corals are often dependent on zooxanthellae for their nutrition.

# Acknowledgements

This book is as much of a surprise to me as it might be to many of you. For that I have to thank Peter Buckman at the Ampersand Agency Ltd for spotting one of my short pieces in *Oxford Today* and being convinced (and convincing me) it could be turned into a book. I must also express my gratitude to Alex Clarke of Headline/Wildfire who immediately got the vision of *The Deep* and had the faith to commission me to write it. Three further heroes in this tale are Celine Kelly, Shoaib Rokadiya and Julia Bruce who put a great deal of work into herding my catlike wandering prose into some semblance of a coherent book. Many thanks to all of you and to all at Wildfire who have also helped to bring this project to fruition (I love the cover too!). There are also, many, many people who have, in one sense or another, shaped my passion for the ocean, my science and many of the views expressed in this book. Without them this story could never have happened and never been told. My grandfather, a true old man of the sea, and my grandmother, who opened a window on another world, the ocean, and the strangeness of its many inhabitants. My uncles

# Acknowledgements

also played their role, granting me the liberty of accompanying them on many a sunny day's fishing. My mother and father were always there, encouraging me to explore the natural world and to follow my dream of becoming a marine biologist. They always believed I could do anything I wanted, but most importantly, they made me believe it was possible too. What more can a child ask of his parents?

Then there are my scientific parents, two of whom deserve a very special mention, Ray Gibson and Paul Tyler, with whom I have shared the joy of discovery of some of the strangest critters on Earth (or should I call the planet Ocean) and to whom I will always be indebted for inspiring me and getting me through the rough patches as well the fantastic ones. There are many others from whom I have learned so much and who have been great workmates and shipmates, some no longer with us: John Gage, Alan and Eve Southward, Steve Hawkins, Quentin Bone, Lisa Levin, Fred Grassle, David Billet, Andrew Clarke, John Croxall, Malcolm Clark, John Lambshead and many others. I must also give a special thanks to the explorers Oliver Steeds, Nigel Winser and Rupert Grey who created the Nekton Foundation, which has trailblazed a new wave of scientific exploration of ocean. Thanks to you and to Lucy, Paris, Molly, Belinda and Catherine, as well as XL Catlin and Jeff Willner of Kensington Tours, among others, for making that first mission to Bermuda possible. A big hug must go to all my students and postdocs who have shared my enthusiasm for the ocean and who I hope will take up the torch of knowledge and make a difference to the ocean. Some of you appear in these pages in some of my most unforgettable moments. I must also give a big shout-out to those who give everything to fight against those who would leave this world a wasteland to enrich themselves. They are a constant voice of sanity when humankind seems set on self-destruction. Mirella von

Lindenfells, Sophie Hulme, Kristina Gjerde, Carl Gustav Lundin, Matt Gianni, Richard Page, Debbie Tripley, Simon Reddy, Dan Laffoley, Chris Reid, Ove Hoegh-Guldberg, Craig Downs, Duncan Currie, David Stone, David Vousden, Peter Auster and Les Watling. You have been the equivalent of the Saxon Shield Wall against the forces of Mordor.

I reserve the final word for my wife, Candida, and my daughters, Zoe and Freya. *The Deep* was really written for you. It is a song for the ocean, one I hope infects its readers with the joy I feel for the life within it. I hope with all my heart that it wakes people up to the perilous situation the ocean is in. By our collective action we can help choose the right road and ensure the ocean remains to support you and fill you and future generations with wonder at its wild and uncompromising beauty.